CHASING HUBBLE'S SHADOWS

CHASING HUBBLE'S SHADOWS

THE SEARCH FOR GALAXIES

AT THE EDGE OF TIME

JEFF KANIPE

HILL AND WANG

A DIVISION OF FARRAR, STRAUS AND GIROUX

NEW YORK

Hill and Wang
A division of Farrar, Straus and Giroux
19 Union Square West, New York 10003

Library of Congress Cataloging-in-Publication Data
Kanipe, Jeff, 1953–
    Chasing Hubble's shadows : the search for galaxies at the
edge of time / Jeff Kanipe.— 1st ed.
        p.   cm.
    Includes bibliographical references and index.
    ISBN-13: 978-0-8090-3406-2 (alk. paper)
    ISBN-10: 0-8090-3406-9 (alk. paper)
    1. Outer space—Exploration.   2. Hubble Space
Telescope (Spacecraft).   3. Large astronomical telescopes.
4. Galaxies.   I. Title.

QB500.262.K36 2006
523.1'12—dc22

                                    2005009652

Designed by Jonathan D. Lippincott

www.fsgbooks.com

1   3   5   7   9   10   8   6   4   2

FOR ALEX

Eventually, we reach the dim boundary—the utmost limits of our telescopes. There we measure shadows, and we search among ghostly errors of measurement for landmarks that are scarcely more substantial.

—Edwin Hubble

CONTENTS

CHASING HUBBLE'S SHADOWS

In the midst of the word he was trying to say,
In the midst of his laughter and glee,
He had softly and suddenly vanished away—
For the Snark was a Boojum, you see.

—Lewis Carroll, "The Hunting of the Snark"

"The cup of the Big Dipper contains over four hundred galaxies," said the planetarium director, emphasizing his remark by rapidly circling the famous star grouping with his red pointer. "Just imagine," he continued, "if there are that many galaxies confined to just this tiny area, then there must be hundreds of millions of galaxies scattered across the entire sky, each containing tens of billions of stars." And with that, he turned a dial on his console, and the stars began to march majestically across the sky to the opening strains of Henry Mancini's "Theme from *A Summer Place*."

Fittingly, it was late summer 1968 and the first time I had ever set foot in a planetarium. My youthful interest in astronomy was still rudimentary, but already I had learned and could recognize on sight most of the constellations. I watched the familiar patterns rise in the east, drift overhead, and disappear in the west as the impossibly high strings of "Summer Place" soared in the background. Perhaps I am easily amused, but I was wowed.

Thirty-six years later, with the public release of the Hubble Ultra Deep Field (HUDF)—the deepest, farthest look into the universe ever made—my first-time planetarium experience came back to me and I was wowed all over again. Astronomers at the Space Telescope Science Institute in Baltimore announced that a region of sky just three thousandths the size of the cup of the Big Dipper contains at least ten thousand galaxies! That's an area you could easily cover with a pencil eraser held at arm's length. Assuming that the distribution of galaxies is similar across the sky, the number of galaxies embraced by the Big Dipper's cup would be well over 190 million, counting all those down to the faintest ones seen in the HUDF. If you extrapolate these values to the number of potentially observable galaxies across the entire sky, you get a truly mind-boggling number: 127 *billion*! Of course, the true population, consisting of these galaxies plus those that cannot be, or have yet to be, detected with current telescopes, would be far greater.

We know how deep Earth's oceans are. We know the width and breadth of the solar system. The size of the Milky Way is no longer in question; nor is the expanse of space lying between the nearby galaxies. But just how deep is the universe? Such a question has the most profound implications, because in this case, *deep* implies more than size and distance on the grandest of scales—it implies age as well. Some of the galaxies in the Hubble Ultra Deep Field, for example, lie more than 12 billion light-years away. But that immensity also represents a span of time—lookback time—of more than 12 billion years in the past, because it has taken that long for the photons from these remotest known galaxies to reach Earth, despite zipping along at a velocity of 300,000 kilometers per second. Hence the universe is far richer than people thought it was in the 1960s or, for that matter, in the three decades that followed. And today the answer still eludes astronomers. Not even the acute vision of the Hubble Space Telescope, the instrument used to make the remarkable HUDF image, can survey such realms. Still, the Hubble telescope has afforded us the first glimpse of the objects that live near the farthest reaches of the universe, billions of newly formed and inchoate galaxies. Only that telescope, at this time in history, could have provided humankind with such a penetrating look at what creation hath wrought.

What a marvel and scientific triumph the Hubble telescope has been. No other contrivance save Galileo's humble two-inch telescope ("Old Discoverer," he called his fifth and favorite model) has done more to advance humanity's understanding of the universe in so brief a time. Even with a modest 2.4-meter mirror, the telescope's location above the distorting effects of Earth's atmosphere has extended our visual acuity more than a billion times. If fireflies flickered in the dark recesses of the moon, we could see them with the Hubble telescope. Its abundant light-gathering power has enabled astronomers to study stars, star clusters, nebulae, and galaxies in far greater detail and at significantly greater distances than anyone had thought possible before it was placed into orbit in 1990. All the more reason to mourn its probable loss due to a federal budget constraint. Stamping out wasteful government spending is certainly a good thing, but wasting an invaluable telescope that has revolutionized the way we look at the universe does not seem to be in the best interest of science or society. But that's another topic for another day.

For the science of cosmology—the study of the origin, evolution, and structure of the universe—the Hubble telescope has been both a pathfinder and a plumb line into the vast depths of our cosmic origins. This was aptly demonstrated in January 1996 with the public release of the Hubble Deep Field, the first baby portrait ever made of the universe. Had it not been for the Deep Field and its complement a year later, the Hubble Deep Field South, we would have never had an Ultra Deep Field. Thankfully, we now have all three.

The original Deep Field was a bold proving ground for later explorations of the distant universe. For ten consecutive days, between December 18 and 28, 1995, astronomers focused the Hubble telescope on a single spot of sky in the constellation Ursa Major (Big Bear). Except for a few very dim stars that were used for guiding purposes, the region initially appeared devoid of galaxies. A total of 342 exposures were made, each ranging between 15 and 40 minutes, covering ultraviolet, optical, and infrared wavelengths. Ostensibly, the purpose of the multiple exposures was to see how many galaxies might be strung out toward the edge of the visible universe and whether any structural features could be discerned in them. Perhaps the image would reveal a few galaxies, or perhaps a few hundred. Maybe the galaxies would appear faint, fuzzy, and featureless or at best show some modest structural detail.

The result was a stunning still life of more than two thousand galaxies, a flurry of budding, tumultuous light whipped up in the shadowy primordial vacuum. This is how the universe looked when it was less than 2 billion years old: very strange; very chaotic. Upon seeing these galaxies for the first time, Robert Williams, then director of the Space Telescope Science Institute and leader of the Deep Field research team, said, "We have not been able to keep from wondering if we might somehow be seeing our own origins in all of this."

When I first sat down with my eight-by-ten copy of the Deep Field, I scrutinized it with a magnifying lens and then I just gawked, trying to fathom what I was seeing. Of course, it's the bright things the eye naturally fixates upon, but in this case my saccadic vision bounced around the photo like a pinball in play. Just left of center was a bright star—a foreground object in our galaxy. It was positioned next to a cream-colored spiral galaxy that probably lay a billion or more light-years beyond. Scattered around the field were nearly two dozen prominent galaxies, most of them spirals or partial spirals with bright, broad centers. Mixed in were a dozen or so ruddy spheroids that stood out like smoldering embers against innumerable smaller shards dappling the background.

Going back to my magnifying glass, I inspected the spaces between the fainter but still obvious "blobjects." Here were flecked wraiths of light that barely surfaced above the darkness—hints, perhaps, of other galaxies. *Hubble's shadows*, I thought to myself. *He*, more than anyone, would not have been surprised to see these features. Indeed, he would have been surprised not to have seen them. In autumn 1935 Edwin Hubble, who until his death in 1953 was the world's preeminent observational astronomer, delivered a series of lectures at Yale University in which he described his principal observations of the universe and their possible implications. The lectures were compiled the following year into a book called *The Realm of the Nebulae*, which is still a must-read for all astronomers because it was largely responsible for presenting the first modern view of the cosmos. At the conclusion Hubble reflected upon the challenges astronomers face in detecting objects at the edge of the observable universe: "Eventually, we reach the dim boundary—the utmost limits of our telescopes. There we measure shadows, and we search among ghostly errors of measurement for landmarks that are scarcely more substantial."

Hubble's murky meditation is usually interpreted as an expression of the challenges astronomers face in observing so vast a universe with limited telescope technology, but for me his words took on new meaning as I looked at the Hubble Deep Field photograph that day in 1996. How overwhelmingly limitless the universe appears, even from the perspective of a powerful telescope unhindered by Earth's palpitating atmosphere. The shadows with which Hubble had to contend were indeed ghostly but not like those cast in the image appropriately bearing his name. Although the field encompasses a mere pinprick in the celestial vault, it is a profound perforation. The largest and brightest galaxies in the image are 7 to 8 billion light-years away, and a few would prove to shine from a universe that existed at least 12 billion years ago, when the universe was about 10 percent its present age. In all, there are at least twenty-five hundred galaxies assorted throughout billions of light-years along the line of sight of this extremely narrow visual tube. Just how many angels can flit through a soda straw?

Observational cosmology allows us to use telescopes as keyholes, through which we spy frozen moments of cosmic lookback time when objects shouted out their existence with heat and light. For nearby galaxies the lookback times range from a few million to hundreds of millions of years, by no means an inconsequential amount. But the lookback times of galaxies in the Hubble Deep Field require that we invert the way we typically denote benchmarks of spatial reference and age in the cosmos. The galaxies in the Deep Field flickered to life so long ago that they are immensely old relative to our present universe. On the other hand, because they are so old, they are *young* relative to the instant of the big bang. Hence we must now reorient ourselves from looking *back* in time, toward the beginning of the universe, to looking *forward* from that beginning until now.

The quest to find the most distant objects in the universe goes back many decades and has proved to be one of science's most challenging and arduous observational undertakings. Researchers had to ask themselves some very tough questions. How did the first stars and galaxies form, and when? What did they look like? How massive and how bright are (or rather, *were*) they? In the twilight years of the twentieth century, there were no ready answers to such questions. All assumptions as to the nature of primordial stars and galaxies were based on the known stars and galaxies in our immediate region of the universe. But trying to link the modern universe to the primordial universe was, and still is, a risky business, presenting astronomers with a kind of cosmological catch-22. Giant galaxies such as the Milky Way, it was assumed, had evolved since the early universe, but the effects of that evolution had to be understood before sound assumptions could be made about the first generation of galaxies. Astronomers knew, of course, that stars like the Sun were products of previous generations of stars that had cooked up atoms of hydrogen and helium in their cores to create the heavier elements—"metals," the astrophysicists call them—upon which life is so dependent. But the origin and nature of the primordial stars and their host galaxies were complete unknowns.

Observations during the mid-twentieth century showed that, on average, all galaxies in the nearby universe, including our own, are very old. Some of the Milky Way's oldest stars, which congregate in dense spherical systems called globular clusters, are about 13 billion years old, almost as old as the universe itself (a fact that once made theorists who held with the standard big-bang model of the universe distinctly uncomfortable). Distant elliptical galaxies that formed when the universe was roughly half its present age also contain very old stars, meaning either that they formed very soon after the big bang or that their stars matured more rapidly than in other galaxies.

By the 1980s it was clear to astronomers that their only hope of finding significant evolutionary differences in galaxies was to observe lookback times that were greater than half the age of the universe, something ground-based telescopes were ill equipped to do. Theoretical models predicted that if and when a population of "primeval galaxies" were found, they would likely look considerably different from the ones in the modern universe. But that was about all astronomers agreed upon. The galaxies' colors, sizes, numbers, and peak formation period remained questionable. As future observations would show, some astronomers' predictions were spot on, while others were considerably off the mark.

"A short version of the story," says Caltech's S. George Djorgovski, a pioneer in the search for primeval galaxies since the mid-1980s, "is that just about every conceivable combination of luminosity, color, size, and so forth was proposed for primeval galaxies over the years, and pretty much all of them corresponded to some formative stage in the evolutionary history of galaxies of some type or other."

A major question clouding the search for primeval galaxies was whether they would be shrouded in dust. Many deep surveys conducted into the early 1990s had failed to turn up any definitive candidates for primeval galaxies. If they were obscured by intervening curtains of dust, that could explain the lack of unambiguous detections. But dust was not something spontaneously created out of the big bang. Rather, dust was a product of generations of stars that had formed enough metals to "pollute" their environments. The first stars, however, were metal-free, or nearly so, meaning that if the first galaxies were obscured by dust, then perhaps they weren't that young after all—or even first.

Some astronomers, Djorgovski among them, speculated that perhaps a few primeval galaxies had already been found. The first such candidate was discovered as early as 1963 by Caltech astronomer Maarten Schmidt. Except for a slight ray of light projecting from it, the object, a radio source known as 3C 273, looked like an ordinary star in the 200-inch Hale Telescope on Palomar Mountain. Its spectrum, however, did not look like any stellar spectrum Schmidt had ever seen. After puzzling over it, he finally realized that the star's spectral lines were displaced over 15 percent from their standard laboratory position toward the red end of the spectrum, a phenomenon known as a redshift. Because the universe is expanding, light coming from sources at progressively greater cosmological distances is stretched like pulled taffy toward longer, or redder, wavelengths. The greater the redshift, the more distant the object and the farther back in time it lies. When applied to 3C 273, Schmidt was astonished to realize that this "star" must lie well over a billion light-years away. How could a star be that bright from that distance? The answer was that it couldn't, unless it was actually closer than its redshift implied and its abnormal spectrum was the result of some intrinsic pathology—an imploded supernova, perhaps, or a supermassive star with an intense gravitational field. Such explanations were quickly ruled out, however, and astronomers had to finally admit that they were dealing with a new type of astrophysical creature.

By the 1970s astronomers conjectured that the intense luminosity of these sources—called quasi-stellar objects, or quasars—was probably sustained by some extremely energetic process going on in their cores, most likely matter funneling into a supermassive black hole. The centers of most giant galaxies, however, are one hundred times less luminous than those of quasars, and if you removed them to any substantial distance, they would rapidly dim.

As such, the possibility stood that quasars and galaxies with active quasarlike nuclei might not represent primeval galaxies as a class.

Another quandary was the size of primeval galaxies. The size of early galaxies was based on two notions of galaxy formation—the "top-down" and "bottom-up" theories. The top-down theory argued that galaxies formed from huge pancake-shaped masses of gas and then broke apart into smaller objects. The bottom-up theory reversed the process: smaller protogalaxies merged to form larger galaxies. Would primeval galaxies, then, appear as large, low-surface-brightness objects, or would they appear smaller and bluer, owing to their first generation of hot stars turning on? Even more uncertain were the roles that initial formation conditions and environment played in shaping the different types of galaxies. Some astronomers thought these galaxies would be highly clustered because they would have initially formed where the density of matter was greatest. Others disagreed, holding that the first galaxies would be isolated fragments, some of which might be undergoing mergers.

One of the biggest sticking points, which didn't pique astronomers until the 1970s, was the role of dark matter in galaxy building. The idea that a galaxy's major component consisted of nonluminous matter whose nature was unknown blurred not only astronomers' ideas about what constitutes a galaxy but also the concept of galaxy formation itself. In a sense it was analogous to the chicken and the egg. If stars and galaxies were tracers of dark matter reservoirs, then how much of galaxy formation was dependent on the formation of dark matter and how much of it was dependent on how normal matter condensed within dark matter halos?

The shapes of primeval galaxies were the cause of great cosmological concern throughout the 1980s. Would one even be recognizable if it swam into plain view of a telescope? What would it look like? A spheroid? A disk? A combination thereof? Disks are dynamically fragile. Mergers and collisions would disrupt their orderly rotation, like tossing marshmallows into a swirling cup of hot cocoa. Would they appear clumpy and amorphous? They would if, as astronomers theorized, galaxy formation was largely a process of building up little fragments into bigger objects. Bright individual knots—subgalaxies—should be apparent in that case.

Astronomers further wondered if there might be a relationship between distance and color or luminosity. Perhaps primeval galaxies went through different brightness phases at different redshift distances, flaring up as their first generation of massive hot stars turned on and then, a few hundred million years later, flickering out as they died off. In the early 1990s Djorgovski and others presciently predicted that most primeval galaxies would be found in a redshift range of 3 to 5, or 1 to 2 billion years after the big bang. And then there were those researchers who thought primevals would be evident largely around a redshift of 3, and that at higher redshifts, corresponding to earlier times, their numbers would drop off sharply, because the primordial gas would still be in the process of condensing.

The only way to know for certain was to go out and look for them, but this would require innovative observing techniques. Because of their great distances, it was generally assumed that these galaxies would be exceedingly faint, and that telescopes and electronic detectors would have to be pushed to the limit if there was to be any hope of finding them.

Intriguingly, one search technique was first used in 1937 by Fritz Zwicky at Mount Wilson Observatory. Zwicky, Swiss by birth, was a colorful eccentric known for a number of unconventional ideas, some of which later turned out to be not so unconventional after all. One of these was that nearby galaxies could be used as gravitational lenses to detect more distant galaxies that lay along the line of sight. This phenomenon, argued Zwicky, was simply in accordance with Einstein's general theory of relativity. The image of a background galaxy, he wrote in *Physical Review Letters* that year, would appear as a luminous ring, its brightness enhanced by the gravitational effects of the intervening galaxy. The discovery of such a lensed galaxy, Zwicky wrote, "would enable us to see nebulae at distances greater than those ordinarily reached by even the greatest telescopes. Any such extension of the known parts of the universe promises to throw very welcome new light on a number of cosmological problems."

Unfortunately, for the rest of his life his appeals to astronomers to look for gravitational lenses fell on deaf ears. A few years later Zwicky had another apparently wacky idea. He found that galaxies in clusters were moving faster than they should, based on the collective mass of the cluster. The galaxies should have flown apart long ago. What was holding them together? Zwicky concluded that it was something called "dark matter." Ironically, his death in 1974 occurred just as astronomers were realizing that a great deal of mass in the universe indeed seemed to be "missing." Five years after that the first gravitational lens was discovered.

Had astronomers given credence to Zwicky's ideas, it's doubtful that much would have come in the way of distant galaxy research. Even with today's telescope technology, gravitational lenses are rare and notoriously tricky to probe. Besides, by the mid-1970s and early 1980s, a new generation of sensitive light detectors were coming online at most major observatories, which precluded such uncertain ventures. CCD imagers enabled telescopes to reach deeper into the universe than was possible with conventional photographic plates and over a wider range of wavelengths as well.

Feeble photons now could be electronically pumped up, bringing fainter objects to the surface. More than any other innovation, it was the advent of CCDs that spurred the quest for primeval galaxies.

Even so, hunting for high-redshift galaxies remained a considerable challenge. A number of attempts were made with ground-based telescopes, but results were often inconclusive. However, in 1985 the first non-quasar, non–gravitationally lensed galaxy beyond a redshift of 3 was discovered. The team, led by Djorgovski, also consisted of Hyron Spinrad, Patrick McCarthy, and Michael Strauss. Up until that time no galaxy had ever been detected above a redshift of 1.82.

A heroic 1988 survey by Anthony J. Tyson of Lucent Laboratories (formerly Bell Laboratories) and a team of Canadian researchers came closest to convincing astronomers that primeval galaxies were within reach. Using the 4-meter telescope at Cerro Tololo Inter-American Observatory near La Serena, Chile, and the 3.6-meter Canada-France-Hawaii Telescope on Mauna Kea, Hawaii, the team made deep images of twelve small fields of apparently empty sky—shades of the Hubble Deep Fields to come. The exposures in visual, red, and infrared colors, each nearly an hour and a half in length, revealed myriad objects—twenty-five thousand galaxies, to be precise—with the bluish colors and luminosities that some astronomers expected of primeval galaxies.

The discovery was widely reported in the science media. In my 1988 *Astronomy* magazine interview with Tyson, he said, "These galaxies are resolved. They are a few arcseconds across, but they are definitely nucleated objects. They are not completely diffuse. There are so many of them that no matter where you point your telescope, you see thousands of them."

Although sensitive CCDs could record their presence, they

were too faint for conventional spectrographs to register their spectra, which could be used to obtain distances. For all anyone knew, these blobjects could simply be a previously undiscovered population of faint dwarf galaxies swarming the local universe, or perhaps modestly bright galaxies at intermediate distances.

Within a decade astronomers were able to identify Tyson's objects as a class of faint blue galaxies that lie at redshifts between 0.5 and 1. Observational evidence suggested that they are small, low-mass galaxies that become visible only when they experience a short burst of star formation, after which they vanish, either by fading or by merging with other galaxies. During their starburst phase, their colors are bluer than any of the bluest galaxies in the local universe, and thus they are clearly not associated with the present population of galaxies. In the February 10, 1996, issue of the *Astrophysical Journal*, Arif Babul of New York University and Henry C. Ferguson of the Space Telescope Science Institute whimsically referred to these objects as "boojums," or "blue objects observed just undergoing moderate starburst." The reference was probably lost on anyone unfamiliar with the Lewis Carroll poem "The Hunting of the Snark." As one of the poem's characters unfortunately discovers, if you found a snark that was a boojum, you would "softly and suddenly vanish away and never be met with again!"

At any rate, whether you call them snarks, boojums, blue galaxies, or blobjects, these bursting dwarfs implied that galaxies had at least undergone modest color evolution since a redshift of 1, implying that they had already formed a generation or two of stars. They were just too blue and bright in the distant past for them not to have undergone some development. Tyson was adamant about what he and his colleagues were seeing, stating in a 1988 issue of the journal *Science* that "we are definitely seeing evolution."

This was probably the first time evidence of cosmic evolution was proclaimed so emphatically in the public arena by an astronomer. There would follow many such declarations as the search for primeval galaxies proceeded full bore. Some claims would turn out to be almost right, others less so.

The first major observational confirmation of primeval, or early, galaxies was the Hubble telescope's "medium" deep field, conducted in 1994. Predecessor to the Deep Field, the images showed a remote cluster of galaxies undergoing some sort of spasmodic turmoil. The distorted arcs and plumes of some of its members strongly indicated violent collisions. Some astronomers proclaimed that this was direct visual evidence that the universe evolves as it ages.

Such enthusiasm, however, did not translate equally throughout the astronomical community. Many of the objects looked peculiar, to be sure, but some looked like normal spirals and ellipticals. Whether the oddballs represented an undiscovered shore of primeval galaxies or just a galaxy group disturbed by gravitational interactions remained to be seen.

All doubt was swept away by the Hubble Deep Field. Not only did it reveal a new population of galactic "snarks," it did so in stunning detail. Suddenly, legions of weird, never-before-seen galaxies were loosed upon the astronomical community.

The bounty found in the Hubble Deep Field was particularly gratifying for Robert Williams, then the director of the Space Telescope Science Institute. For one thing, it launched the careers of a number of his young team members who have gone on to make names for themselves in early galaxy research (some of whom appear in this book). For another, Williams had stuck his neck out to support the project despite serious criticism and doubt from some quarters and predictions that it would be a complete failure.

Though not widely known, these uncertainties nearly scuttled the Deep Field project before it began.

Six years before the Hubble telescope had even been launched, a contingent of prominent astronomers led by astrophysicist John Bahcall argued that the long exposures would not only consume an inordinate amount of telescope time—involving more than 150 orbits—but that the results were likely to be disappointing. Their tests suggested that the telescope was incapable of revealing a new population of galaxies, which would be too far away and too dim to be seen.

The sobering conclusions had plenty of merit. Galaxies are characteristically bright in the middle, where most of the stars are concentrated, and fade rapidly toward the less-populated edges. As any amateur astronomer knows, only the brightest galaxies in the nearby universe have any visual presence to speak of. To see them in moderate detail requires integrated photographic or digital camera exposures. With more distant galaxies, only the largest telescopes are capable of coaxing out subtle structural features. Not only would galaxies at cosmological distances be smaller and fainter still, but their light would be stretched by expansion well out of the visible part of the spectrum and into the dimmer red sectors. The astronomers argued that this diminution-with-distance, known as cosmological dimming, could very well sap the visual presence of the really distant galaxies. All that would be seen of them, if anything, would be their hazy central regions and little or no structure to speak of.

Conversely, another concern brought up by astronomers was that perhaps too many objects would turn up in the field, resulting in a mass of indistinct and overlapping galaxy gobbets, an optical threshold that astronomers refer to as the "confusion limit." Such was the case with Tyson's galaxies in 1988. Either way, the Deep

Field could turn out to be a big blunder and an embarrassment both to the Space Telescope Science Institute and to NASA.

Williams, an unassuming, soft-spoken man who generally shies away from the limelight, may have also been a little worried that he might end up looking like Geraldo Rivera opening Al Capone's safe only to find a few empty whiskey bottles. Happily, it turned out that the skeptics had overestimated the effects of both cosmological dimming and overcrowding, and Williams and the Deep Field team were spectacularly vindicated.

Today Bahcall is more than happy that his group's predictions turned out to be off the mark. "I was delighted that nature had once again surprised us," he recalls, "and thrilled that the repaired Hubble telescope had the acuity to see the earliest galaxies. The Hubble Deep Field is such a gold mine of information about the history of the universe that astronomers of all types worked together using observations of all kinds to piece together the intricate story of how large-scale structure, galaxies, stars, and ultimately people formed."

Indeed, the HDF's impact on cosmology in the ensuing years has been nothing short of revolutionary. For one thing, it has stimulated an unparalleled amount of follow-up research. In 2004 it was ranked first out of the ten most productive Hubble programs and second in the number of citations made by other researchers. For another, it liberated astronomy from the often unproductive and at times contentious practice of research isolationism, in which astronomers pursue their own investigations and objects without sharing results or pooling resources.

Williams is particularly proud of this accomplishment. "I decided to make the HDF data available to everyone," he says. "Until that time everyone kept their important scientific results in drawers that were closed and locked. It was just the culture. There is a basis for it called intellectual property rights, and that's a very serious thing. But on the other hand, I thought that it had gone too far. It was important for this institute to do something that was groundbreaking and on behalf of the community."

Williams, practiced in the art of seeing the big picture, was confident that outside researchers would respond similarly, and he was right. "That changed *everything*," he recalled. "Keck Observatory got the redshifts [for the Deep Field galaxies]. Until that time you couldn't get a redshift out of those people to save your soul! So we did it for the public, and the other researchers followed suit. It was a sea change that has had as profound an impact on science as the Deep Field itself."

The Hubble Deep Field was a new beginning in cosmological research. The momentum of observations and discoveries since then has propelled the science into a new era of confidence. The rudimentary objects sprinkled across the various deep fields reveal a wealth of knowledge about conditions in the early universe, conditions that also tell us about how the universe is currently evolving.

On the other hand, the Deep Field also hit the reset button on astronomers' previous conceptions about primeval galaxies. "Now I'd say the meaning of 'primeval galaxy' is less clear than ever," says Djorgovski. "It has evolved a lot over the years. If one would say something sufficiently vague to be almost meaningful, such as 'galaxies in the very early stages of formation . . . ,' then you have to say, 'what do you mean by *very early*?' The first X years? Y percent of the present age? The epoch when the first Z percent of mass or light was assembled? And by *formation*, do you mean mass assembly, initial collapse of the halo, the first stars forming? None of the simple, clean-cut definitions [about primeval galaxies] were right. Nature turned out to be far more complex."

Today most astronomers sidestep the semantic sand trap of words like *primeval* and *primordial* and just call the objects young or high-redshift galaxies (though I'm rather fond of snarks and boojums). As the distinctions between the physical properties of these objects become clearer, so too will our cosmic lineage. It is from these preexistent realms that the universe we live in emerged. To borrow from Francis Bacon's reference to light, these galaxies are God's first creatures, the cosmic DNA from which all other galaxies were patterned. The congenital attributes of these objects provide astronomers not only a blueprint for cosmic structure but also a diagram of our origins.

Says Mario Livio, senior scientist at the Space Telescope Science Institute: "From the deep fields we have learned of the cosmic star-formation history—namely, the global rate at which the universe was forming stars as a function of time. This is not only important in itself (being a sort of an 'executive summary' of the history of the universe) but also has direct implications for the emergence of carbon-based life in the universe. For example, from the cosmic star-formation rate one may calculate the rate of carbon production in the universe. You could argue that the most likely time in the history of our universe for carbon-based life to have emerged (at least in large numbers) is somewhere near the peak in the carbon-production rate."

In his opening remarks at the unveiling of the Hubble Ultra Deep Field in March 2004, Steven Beckwith, Williams's successor as director of the Space Telescope Science Institute, said, "All of the great cultures have stories about origins. We have a deep need to understand our past—where we came from and where we will go. We are very fortunate to live in a time when we can address some of life's most profound questions of science. When [the HUDF] image is fully studied by the astronomical community, we expect it to reveal the secrets to the origins of stars and galaxies and ultimately to ourselves."

The Hubble Deep Field and the Ultra Deep Field stand as two of astronomy's greatest achievements in the quest for cosmic origins, but they are not the only success stories. New orbiting telescopes and innovations in telescope and light-detection technology have empowered astronomers with an exciting array of observational techniques with which to sound the depths of the universe. Deep images made with ground-based telescopes render wider-angle vistas of distant galaxies as well as those that lie at intermediate distances. These observations provide astronomers with a larger representative sample of galaxies than is available in the HUDF and may soon determine whether galaxy evolution progressed differently in regions where there are more galaxies than where there are less. Contributions from telescopes both on the ground and in space that are sensitive to infrared and near-infrared wavelengths uncover the star-formation histories of distant galaxies, showing how they grow. X-ray and radio telescopes reveal information on the gas content of galaxies and the development of black holes and quasars, while also detecting sources that might have illuminated the universe half a billion years after the big bang. Finally, observations at microwave wavelengths are mining the tenuous relict radiation of the big bang itself, the cosmic microwave background. This all-sky glow, the temperature of which is slightly less than 3 degrees above absolute zero, developed at a time before galaxies existed. Measurements of the warmer and cooler furrows in the cosmic background are beginning to reveal where and when the first stars and galaxies likely took root.

With observational tools such as these, as well as the power of computational modeling, cosmologists are, for the first time, stitching the near and far universe together with facts rather than conjectures. Once-fuzzy parameters like the ratio of normal matter to dark matter, the rate of the expansion of the universe, and the effects of that mysterious overlord of cosmic content, dark energy, are being whittled down to precise values. More important, these estimations seem to be in close agreement with analyses made by independent means. In nearly a century of observation and theory, cosmology has never witnessed such an overwhelming consensus.

That is not to say the path is free of obstacles. There are any number of outstanding questions whose possible answers give cosmologists nightmares. What if dark matter cannot be detected? How does one account for an accelerating universe? And what if more galaxies with mature stars are seen beyond those already considered to be the youngest in the universe, like those in the HUDF? How could such a rapid assembly of galaxies be explained?

Such challenges, formidable though they may be, are in fact cosmological catalysts, inspiring innovative ideas and methodologies, defining objectives, and in general propelling the science toward greater insights. Whatever else might be said, it is clear that cosmology is going places. In the following decades research promises to be as heady as it was in Edwin Hubble's day and as illuminating as in the days when Galileo first turned his telescope toward the heavens. Astronomers may still be chasing shadows, but they are beginning to glimpse the objects that cast those shadows. Our eyes are finally adapting to the dark depths of the universe, the place of our origins, where there is more light than can be imagined.

THE RACE FOR THE MOST DISTANT GALAXIES

It is a very human tendency to climb a celestial mountain. So it stands for any race, including that of finding individual objects at greater and greater distances.

—Hyron Spinrad, *Astrophysics Update*, 2004

Jean-Gabriel Cuby knew he was in Hawaii, but where he was standing was a paradise of a different sort: an astronomer's paradise. The temperature was in the upper thirties, and the gravelly ground was well dusted with snow. Parkas, gloves, and boots were de rigueur, not aloha shirts and flip-flops. The fragrant, carefree beaches of Hilo and the Hamakua Coast were miles away and nearly 14,000 feet below this place, the Mars-like summit of Mauna Kea, one of the Big Island's premier dormant volcanoes.

Situated above 40 percent of Earth's atmosphere, this rarefied location may as well *be* Mars. The air is thin and cold. Astronomers and tourists alike have been known to pass out from lack of oxygen or, more commonly, to experience muddled thinking from the same effect. Except for an indigenous species of tiny flightless insects called wekiu bugs, which feed on hapless dead and dying

arthropods lofted upslope by surface winds, and the recently introduced wolf spider, the craggy terrain is virtually lifeless. Apollo astronauts might have trained in places like this, and it's no stretch to
imagine it serving as an analog for future Mars missions.

For all these same reasons, it's also the perfect place to conduct research astronomy. Telescopes—and there are over a dozen
of them sprinkled along Mauna Kea's quiescent summit ridge—
have nearly unsullied views of the universe. A tropical inversion
layer well below the summit ensures that moist air at lower alti-
tudes rarely mixes with the drier air at the top. Thus moonless
skies are sable black, the vitreous atmosphere steady, and the viewing conditions ideal or, as astronomers call it, "photometric."

Such were the conditions the night of November 18, 2001.
Cuby was standing on the catwalk encircling the dome of the
3.6-meter Canada-France-Hawaii Telescope (CFHT) looking up
at the stars and counting meteors. It was well past midnight, and
the Leonid meteor shower was at its peak. That year Hawaii was a
particularly favored viewing locale. Cuby estimated he was seeing
about three shooting stars per minute, some with spectacular tails.

He was at Mauna Kea to look not for meteors, however, but for
something more elusive and much, much farther away: a galaxy,
the most distant one he could find. Such a galaxy might prove to be
very young, one perhaps whose nascent stars have just turned on
to light up its dark surroundings. In a universe of tens of billions of
older, more proximal galaxies, such a specimen would indeed be a
pearl of rare price.

It was the first night of his four-night observing run, and both
the sky and the weather were cooperating beautifully. Though near
freezing, the temperature was actually mild for that time of year.
Best of all, the sky was calm, completely clear, and free of excess

humidity. Observing conditions couldn't be better, except in space itself.

Then an unexpected terrestrial threat appeared on the scene in the form of a solitary car, headlights blazing, wending its way along the sinuous summit road. The occupants were undoubtedly hoping to find a unique vantage point from which to watch meteors. The car meandered toward Cuby's location, and for one horrifying moment the headlights flashed across the CFHT dome. From his position on the catwalk Cuby frantically shouted at the unwitting intruders to shut off their lights, but apparently they were too oblivious or oxygen-starved to notice the threat they posed. The car eventually circled around and rolled leisurely back down the mountain, leaving a number of wrathful astronomers in its wake. Fearing the worst, Cuby dashed down to the control room, where he was relieved to see that his 15-minute exposure was undefiled. The computer screen was dappled with galaxies. Apparently the dome had shielded the giant telescope from the unwanted glare.

Notwithstanding the incident of the errant automobile, all three of his subsequent nights at CFHT were clear and free of intrusive light. Each morning he drove back down to Hale Pohaku, the base camp at 9,000 feet where visiting astronomers can stay during their run. Observing on Mauna Kea is a wonderful experience, Cuby says, but dawn on Mauna Kea is his favorite time. "The view is unbelievable. The huge Mauna Loa volcano is in front of you, all clear down to the valley where Saddle Road passes. Incredible clarity. And best of all, the craters of the tens of volcanoes are illuminated horizontally by the Sun. Spectacular! Then you arrive at Hale Pohaku, a little dizzy because of the change of altitude—more oxygen to breathe—and then you sit down in the dining room and you have a good breakfast of eggs and hash browns. I've never been disappointed with such Mauna Kea mornings, because it usually means you've had a successful observing night. Otherwise you go down earlier at night if it is cloudy, and you miss this special time. It's worth a hardworking astronomer's life to get to live this!"

During his four nights Cuby made a total of eighty exposures, or "integrations"—some twenty hours' worth—through a filter that favors a slice of the spectrum located in the near infrared, just beyond the red portion of the electromagnetic spectrum visible to human sight. The observing done, he gathered his data and returned to his headquarters at the European Southern Observatory in Santiago, Chile, where he painstakingly began searching for the telltale characteristics of a very distant and young galaxy. Although he didn't know it at the time, one of his Mauna Kea exposures would prove to be a winner in more ways than one.

The way astronomers observe the universe has changed greatly since the days when they would manually guide big telescopes, make hours-long exposures of the sky, often in numbing cold, and then examine the photographic plates the next afternoon and exclaim "Eureka!" or, more typically, "Damn!" These days astronomers are looking for objects that are some 10 billion times fainter than what can be seen with the unaided eye. To do that effectively, efficient (and expensive) electronic detectors and powerful imaging software have superannuated photographic plates as the recording media of choice. Although the innovations have extended astronomers' reach toward greater distances and fainter objects, mining the rich seam of galaxies extending to the edge of the visible universe requires both time and exactitude. Numbers must be crunched, data reduced to its essence, image sets processed and examined, and statistics weighed. Then everything must be checked and rechecked. Often further observations must be conducted to complete a long-term project or to build on previous observations.

In short, rarely does a eureka moment follow quickly on the heels of the actual observations. Cuby's case was no exception. A little over a year after his CFHT observing run, he had to venture to the Very Large Telescope in northern Chile to make follow-up observations. The VLT consists of an array of four 8.2-meter telescopes located at the summit of 8,644-foot Cerro Paranal in the Atacama Desert, one of the highest, driest places on Earth. One of the VLT's specialties is its ability to separate the light of a very faint galaxy into its component colors, or wavelengths, the result being a narrow band of bright and dark lines called a spectrum. The placement and strength of the spectral lines correlate to the galaxy's chemical composition, but more important, they also attest to its location in space and time. For this observing run, Cuby's job was to "reacquire," or reobserve, some of the galaxies captured during his four nights on Mauna Kea, take their spectra, and determine their distances from Earth.

His preliminary results indicated that one of his galaxy candidates was very far away indeed, lying in a field with a handful of other galaxies that in the CFHT frames appear as little more than pixilated blots. But what set this particular galaxy apart was that it was extremely shy. At visual, blue, and red wavelengths it hid in the sky background, completely invisible. Only when observed at longer wavelengths, in the infrared—specifically, around a wavelength of 920 nanometers (the mean visual wavelength is 550 nanometers)—did the galaxy blushingly reveal itself.

Cuby planned to winnow the light passing through this narrow breach and look for a spectral feature christened the Lyman-alpha line. For galaxy hunters, this is one of the most information-rich spectral lines in the entire electromagnetic spectrum. It arises from the radiation output of a galaxy's population of energetic hot, young stars, and thus it imparts clues about the system's stellar content. But it's much more than that: because its light is not seen at wavelengths shorter than infrared, its presence is also the spectral semaphore of a distant galaxy.

In recent years astronomers have used the Lyman-alpha line with great success to track down very distant galaxies. The key to this success, of course, is having a telescope with the light-gathering power required to produce the spectrum of so faint an object. But—and this is no small detail—it's also important to know where to find Lyman-alpha in a particular galaxy's spectrum.

In laboratories the Lyman-alpha line can be found at its usual wavelength address of 121.5 nanometers, which actually places it in the ultraviolet part of the neighborhood—too short to be detected visually. But put it into an expanding universe, and like the drone of a receding airplane, its wavelength shifts downward from the ultraviolet, through the visual, and into the red and infrared wavebands. Moreover, most if not all of the emission at wavelengths shorter (or blueward) of where the Lyman-alpha line peaks will be sharply cut off due to absorption by hydrogen gas present in the galaxy's own interstellar medium. When the spectrum is converted into a line profile, where peaks represent emission and valleys represent absorption, the Lyman-alpha line juts out, looking like the sharp crest of a tall wave. The wave marches from the ultraviolet band in galaxies at distances of about 11.5 billion light-years, through consecutively longer, redder wavelengths, out to galaxies well beyond 12 billion light-years.

Since Cuby's shy galaxy was emitting radiation strongly at a wavelength of 920 nanometers, he assumed the Lyman-alpha line should be in the vicinity as well. But—and this was his first surprise—it wasn't. Puzzled, he began a wider visual analysis of the spectrum and found a strong, distinct line with the characteristics he was looking for, but at a shorter wavelength—872 nanometers. An Italian colleague who was looking over his shoulder in the VLT control room was convinced it was the stray Lyman-alpha line, but Cuby wasn't so sure. When he returned the next day, he repeated his analysis and got the same results. Indeed, it *was* the Lyman-alpha line, with a redshift of 6.17, which corresponded to a look-back time of 93.3 percent of the age of the universe. This galaxy was well over 12.5 billion light-years away in a universe only 930 million years old.

That in and of itself was an extraordinary find, but there was more: just redward from the galaxy's Lyman-alpha line was a faint smudge of emission called a continuum, in which no discrete lines were apparent. This was a second, but significant, unexpected re-sult. The source of a continuous spectrum can be a solid, a liquid, or a gas, but rather than emitting selectively, it emits across the board at *all* wavelengths. A lamp filament, for example, produces a continuous spectrum. So does the Sun, and so do other stars.

The smear of light downstream from the Lyman-alpha line in Cuby's galaxy was the result of a gang of ultraviolet sources—hundreds of millions, perhaps billions, of stars. The reason this galaxy is bright is because it is illuminated by its own stellar population, one that must have formed several hundred million years *earlier*. In its brief lifetime, this galaxy has matured in a universe that itself has barely taken its first steps.

Galaxies with redshifts of 5 and 6 are no longer as rare a find as they were in the late 1990s. In 2004, astronomers announced the first crop of redshift 7 and 8 galaxies, and now they are promising to find more, perhaps some at still greater redshifts. Most, like Cuby's, were detected by their Lyman-alpha emission. The most recent search strategy is actually an incarnation of Fritz Zwicky's original idea, whereby a massive distant cluster is used like a magnifying lens to see galaxies that lie farther behind it. If nearby galaxy clusters are massive enough, their gravitational fields can bend light rays emitted by background sources, amplifying them by as much as a hundred times. Combined with the light-gathering power of 8- and 10-meter-class telescopes, spectra can be taken of these lensed sources and their distances determined. One such galaxy, discovered this way by astronomers in the Southern Hemisphere in early 2004, was thought to have a redshift of 10. If true, this would place it in a universe only 470 million years old. Its stars would have had to coalesce at least 100 million to 200 million years before that, when the universe was still enveloped in a pall of neutral hydrogen, the murky miasma of the big bang's aftermath. As of this writing, however, the redshift value for this galaxy is still in question.

The protean objects stippling deep surveys contain a wealth of information about conditions in the early universe. Their numbers, structures, and colors directly relate to the degree of crowding of galaxies during these epochs, the role environment may have played in their development, and the ages and masses of their component stars. Those same conditions also tell us about how the universe is currently evolving. For the first time, astronomers can look at galaxies *then* and compare them to galaxies *now*. How are they different, and how are they the same?

Sometimes they're remarkably the same. Sangeeta Malhotra, an astronomer at the Space Telescope Science Institute and the principal investigator for the Grism ACS Program for Extragalactic Science, or GRAPES, says that although high-redshift galaxies may look weird visually, "we don't see any special massive stars or very blue stars making up these galaxies in the early epochs of the universe. We can [make] do with stars like those in the modern universe." The GRAPES project, which has afforded researchers the faintest unbiased spectroscopy of the Hubble Ultra Deep Field, has turned up galaxies out to a redshift of 7. "I would bet that when we look at galaxies at redshift 6," Malhotra predicts, "we'll find a mixture of young and old stars like we do in galaxies at redshift 1."

But the *then* and *now* galaxies can also be strikingly different. First of all, the distant galaxies look very strange. Some are bar-shaped with multiple bright regions within; some look like exclamation points and semicolons. They are bi-lobed, tri-lobed, and so train-wrecked that you can't tell if they're products of external collisions or internal strife. By contrast, most of the galaxies in our local cosmos look normal, with symmetrical shapes like spirals, ovals, and spheres. Galaxies in the distant universe are also typically smaller than those in the regional universe, probably because they haven't had the time to build themselves up into the more massive galaxies we see today.

Then and *now* galaxies are particularly distinct when it comes to how prodigious they are—or were—at making stars. Alan Heavens of the Institute for Astronomy at the University of Edinburgh and his colleagues have observed young stars in more than ninety thousand distant galaxies located at different epochs in the past. They concluded that 5 billion years ago the overall star-formation rate began steadily decreasing to the rate we see today. In a Royal Astronomical Society announcement of their findings, Heavens sounded a tone of finality: "Our analysis confirms that the age of star formation is drawing to a close. The number of new stars being formed . . . has been in decline for around 6 billion years—roughly since the time our own Sun came into being."

Which raises the question: Could the Sun be one of the last of its kind? Perhaps. Many of the Sun's ancestors have long since departed this universe, gone the way of stardust and cinders. Their halcyon days occurred very early in the universe, though not everyone agrees just how early. Astronomers participating in a project called the Great Observatories Origins Deep Survey (GOODS), an affiliation of three space telescopes—the Hubble Space Telescope, the Chandra X-ray Observatory, and the Spitzer Space Telescope—favor an early phase of star formation. They claim that the universe was already vigorously producing stars when it was only a billion years old—around a redshift of 6—and continued to do so for the next billion years or so. Therefore the onset of large-scale star formation should have begun even earlier, perhaps as early as 700 to 800 million years after the big bang.

Other research, however, suggests a very different scenario, one in which star formation starts out sluggishly, then undergoes a dramatic evolutionary kick in the pants. A team led by Andrew Bunker of the Institute of Astronomy at Cambridge University—one of the first researchers to analyze and present results based on the Hubble Ultra Deep Field—concluded that the rate of star formation was actually six times greater 2 billion years after the big bang, at a redshift of 3, than it was when the universe was much younger, at a redshift of 6. The difference in redshift translates to only about a billion years, but the difference in star-formation rates is significant, because these earlier stars were likely the progeny of those that lit up the rest of the universe.

Today we can see across billions of light-years of space. But a billion years after the big bang, much of the universe was filled with opaque, neutral hydrogen gas created in the aftermath of the big bang. Stars formed in isolation within murky concentrations called dark matter halos. To be seen from the distance of neighboring stars, they had to burn away, or ionize, their surrounding shell of neutral hydrogen. Eventually the radiation energy of enough stars escaped to irradiate their galactic environs. Subsequent generations of stars would have had the combined radiant capacity to burn off much of the neutral hydrogen fog, making the universe transparent to photons. This process, called reionization, is believed by most astronomers to have begun in earnest around 300 to 400 million years after the big bang, at a redshift of 17, and concluded 500 million years later at a redshift of around 6. Knowing the history of reionization more precisely would help astronomers reconstruct much about the formation processes of the first stars and galaxies, which have in turn evolved into the large-scale structures seen today.

Although astronomers are uncertain as to when reionization began, they are fairly confident about when it came to an end. Perhaps the best evidence was found in 2001 by a team led by Robert Becker, of the University of California at Davis, and Xiaohui Fan, now with the University of Arizona. During their analysis of the spectra of several high-redshift quasars detected by the Sloan Digital Sky Survey, they discovered an absorption feature called the Gunn-Peterson trough. It was the first time such a feature had been observed, though it had been predicted in 1965 by James Gunn of Princeton University, now with the Sloan Digital Sky Survey, and Bruce Peterson of Caltech. As we saw with high-redshift galaxies, interstellar neutral hydrogen absorbs starlight very effectively at wavelengths blueward of Lyman-alpha. But for quasars shining within the reionization epoch, absorption is largely the result of neutral hydrogen present in the intergalactic medium on a line of sight between ourselves and the quasar. If we look at the spectrum of such a quasar, we see a jagged peak of Lyman-alpha emission that, just blueward, abruptly drops off to an almost completely flat plain. This absorption trough represents the neutral hydrogen present in the intergalactic medium before the universe became completely transparent. This is why the effect is a good test for when reionization ended. Quasars with redshifts slightly greater than 6 exhibit a decided Gunn-Peterson trough, while the effect is much less pronounced in quasars with redshifts less than 6. Since this discovery, Fan has found a handful of redshift-6-and-greater quasars with complete Gunn-Peterson troughs, indicating that the tail end of reionization must have occurred about a billion years after the big bang.

But the redshift 6 threshold need not be, and probably is not, an absolute boundary. Independent of Becker and Fan's discovery, and just prior to their 2001 announcement, another team, led by Caltech's George Djorgovski, found that the spectrum of a redshift 5.7 quasar had a "patchy" appearance to it at wavelengths shorter than the Lyman-alpha emission line. Apparently some, but not all, of the quasar's light was being absorbed by clumps of neutral hydrogen existing alongside ionized regions. It was the kind of spectrum, says Djorgovski, that you would expect from a quasar residing on just this side of the reionization epoch. "It is the approach to reionization," he says.

This finding suggests that the end of reionization may have lasted for 100 million years or more, perhaps wrapping up in one sector of the universe while still millions of years away from doing so in another. The distribution of matter and the emergence of large-scale structure would be among the determining factors in how soon reionization ended. The results of galaxy surveys made over a much wider area of sky than the Hubble Ultra Deep Field support this notion. In September 2004 Sangeeta Malhotra and James Rhoads of the Space Telescope Science Institute announced their discovery of a "sheet" of galaxies dispersed across space at a redshift of 5.9—a difference of only 21 million years from a redshift 6. The implication is that reionization would have been more robust in areas where galaxies are densest. The HUDF image frame happens to be located along one edge of this galactic sheet, where galaxies are less dense. The presence of these structures, says Rhoads, would affect the reionization of the universe, "because the ultraviolet light that separated intergalactic hydrogen atoms into protons and electrons would have been more intense where galaxies are more common. It is then likely that reionization proceeded at different speeds in different regions of the early universe."

Whatever its timetable, reionization wouldn't have occurred at all if there weren't enough stars, or time, to do the job. Although it doesn't take long for stars 10 to 100 times the mass of the Sun to form—100,000 to 1 million years or so—once they've started down the main path of their life, such massive stars live only a few million years. This raises a couple of questions. If these stars live and die so rapidly, then how can full-fledged galaxies like the one Cuby found muster themselves inside of 900 million years? Moreover, given that most galaxies and all quasars are thought to contain supermassive black holes in their cores, how do you make a billion-solar-mass black hole in only a few hundred million years? This wouldn't be so problematical if there were only a few such galaxies with high redshifts, but some astronomers fret that a whole lot more may be dwelling in a universe less than 700 million years old.

Indeed, in February 2004 a Caltech-based group led by Jean-Paul Kneib and Richard Ellis, the latter a veteran of high-redshift galaxy searches, announced the discovery of a redshift 6.8 galaxy via the gravitational lensing technique. Light from the galaxy passed near a massive cluster of galaxies on its way to Earth and was magnified about 25 times by the cluster's collective gravity. Following that discovery, a University of Arizona team led by Eiichi Egami observed the same galaxy using the infrared-sensitive

Spitzer Space Telescope. The Spitzer observation is important because infrared detectors are especially sensitive to the presence of older, more established stars in distant galaxies. Using the combined Hubble and Spitzer observations, astronomers were able to determine both the age and the mass of one of the most distant galaxies known.

With a redshift of 6.8, the galaxy exists in a universe that is already about 800 million years old. The presence of a considerable population of older stars within the galaxy, however, indicates that this system has been steadily generating stars for as long as 200 million years. Thus it must have achieved galaxy status at least 600 million years after the big bang, at a redshift of 9.

"The mass of this object is quite respectable," says Ellis, who places it in the 300-to-500-million solar mass category. "It's comparable to a dwarf galaxy, much larger than we'd expect [for this cosmological epoch]. The area that we scanned to find this object was tiny. So unless we've been extremely lucky, then this kind of object is quite common."

Meaning there may be a lot more nascent, star-forming galaxies out there that came together when the universe was only a few hundred million years old. Like mild-mannered Clark Kent fleetly transforming himself into Superman inside a phone booth, galaxies would have had to pull themselves together in a cosmic wink—a few hundred million years. Is that possible?

The answer, say most astronomers, may lie in the nature of dark matter, the chimeric material that is said to make up nearly a quarter of the universe but that, so far, has eluded direct detection. In a young universe that was a thousand times smaller and denser than the one we have now, everything was closer together. Dark matter could have provided the additional gravitational draw to collapse normal matter rapidly into stars and galaxies. In fact, most astronomers tend to agree that dark matter was normal matter's "silent partner" in the galaxy-formation business. "The collapse of early objects is one of the safest predictions the theorists have been working on," says a confident Ellis. "Dark matter could do it."

Indeed, recent observations with the Chandra X-ray Observatory have provided direct evidence that a distant quasar (of redshift 6) containing a supermassive black hole formed less than a billion years after the big bang. If black holes on the order of a billion times the mass of the Sun can exist that early, it suggests that they formed much earlier, as did their host galaxies. At present there are a dozen quasars with redshifts around 6. Obviously, the thing to do now is to probe for quasars at even higher redshifts to determine how quasar properties like the disk and central black hole engine evolve with time.

While astronomers in general are optimistic that answers will be found to such structural timeline problems, some believe that they do not yet have enough facts to draw hard conclusions. "The fact that we don't know what 95 percent of the matter-energy density of the universe is tells us that we're not even scratching the surface of what's actually out there," says astronomer Eric Bell of the Max Planck Institute for Astronomy in Heidelberg and a researcher with the COMBO (Classifying Objects by Medium-Band Observations) galaxy survey. "We have no clue about the physics that drives all this. And that's great! It would be total arrogance for us to imagine that, after four hundred years of telescopes and after only fifty years of getting outside the Earth's atmosphere to do astronomy, we could answer all the questions."

Despite lingering uncertainties, astronomers involved in the search for the youngest galaxies in the universe exude a great deal of consensus and confidence. This attitude is evident in the abundant superlatives offered by the astronomers themselves in popular articles and in the typically reserved scientific journals. Phrases appear like "we are making incredible advances in our knowledge of the early universe"; "cosmology has entered upon its most exciting period of discovery yet"; and "new findings are pouring in." Of course, one still encounters the "we have our work cut out for us" caveat, but such statements are not intended to take researchers down a notch. Rather, they serve as a kind of call-to-arms for observers and theorists alike to prepare for a coming cosmic resurgence. Even the cautionary phrases are based on great expectations.

Indeed, a kind of renaissance *is* occurring in observational cosmology. Given improved telescopes and instruments, says Cuby, "observations of remote and faint galaxies have become possible that were, until recently, astronomers' dreams." But as any history buff will tell you, before the stuff of dreams can be conjured in the Renaissance, one must first pass through the Dark Ages.

TO THE END OF
THE BEGINNING

Or play the game existence to the end,
Of the beginning, of the beginning . . .

— John Lennon and Paul McCartney, "Tomorrow Never Knows"

Rodger Thompson is holding court with a group of reporters before a wall-sized blowup of the Hubble Ultra Deep Field, successor to the Hubble Deep Field as the deepest portrait of the universe ever made. He is dapper in coat and tie, his eyes twinkling behind large gold-rimmed glasses. Like Buddha, he wears the faintest of smiles. It's March 9, 2004, during a morning break in the press conference announcing the public release of the HUDF image and data set. Like other astronomers there, he is in a celebratory mood, despite NASA's recent cancellation of a repair mission to the Hubble telescope, which Maryland senator Barbara Mikulski had roundly denounced from the podium of the Space Telescope Science Institute's conference room just an hour before. Uncertain though the telescope's future is at the moment, the Hubble Ultra Deep Field release is cause for community-wide jubilance, as is apparent by the excited jabber in the room between reporters and scientists. But Thompson has even more cause for rejoicing.

An astronomer with the University of Arizona's Steward Observatory, Thompson is the principal investigator for the telescope's Near Infrared Camera and Multi-Object Spectrometer, or NICMOS. Between September and November 2003, while Hubble's Advanced Camera for Surveys (ACS) was beginning its series of exposure runs on the tiny speck of sky that would ultimately bloom with galaxies, NICMOS was concentrating on a region nested within the UDF frame. Although slightly smaller in scale, the NICMOS field bores deeper into the universe than the ACS. Whereas the ACS image shows galaxies that existed 750 million years after the big bang, NICMOS can potentially reveal galaxies that existed when the universe was a mere 300 million years old. That's one of the reasons Thompson is feeling all right with the world this day. His team has labored to present the NICMOS data and image at the same time as the HUDF, and they have succeeded.

The NICMOS advantage lies in its ability to image galaxies at infrared wavelengths. Infrared telescopes are used to examine the nearby universe, to peer into dense clouds of interstellar dust harboring hot, young stars, or to study the internal features of active galaxies. At cosmological distances, the situation is quite different. The light of galaxies is reddened less by intervening dust than by the expansion of the universe. Expansion stretches visual and ultraviolet light toward longer and longer wavelengths. The greater the redshift value ascribed to a particular galaxy, the farther away it has been carried on expansion and the redder it will appear. No wonder, then, that faint red blobjects like those in the HUDF and NICMOS images make such promising candidates for very-high-redshift galaxies.

During his presentation earlier that morning, Thompson had stressed this very point. The reddest galaxies in the Hubble Ultra Deep Field lie at redshifts of around 6, corresponding to a time about 750 million years after the big bang, he said. But NICMOS can go even deeper. "The implication," he said, "is that the [NICMOS] image has the sensitivity to see standard bright galaxies out to a redshift a little greater than twelve, and that's when the universe was only about 300 million years old, or only 2.4 percent of its present age. This image represents the furthest lookback time for objects ever achieved."

To observe the highly redshifted light of galaxies over such space-stretched distances, the infrared sensors aboard NICMOS must be chilled almost to absolute zero so that heat from the camera doesn't interfere with the detection of galaxies just emerging from a darkling universe. NICMOS was built to operate at temperatures so frigid that even the remote "warmth" of galaxies billions of light-years away would register with its detectors. Ironically, though, the strength of its design was very nearly the cause of its demise.

NICMOS is essentially a can of sensors inserted into a larger can filled with solid frozen nitrogen. This can-within-a-can configuration, called a dewar, allows the detectors to be continuously re-cooled to a stable temperature 60 to 70 degrees above absolute zero. During extensive ground testing in 1996, engineers allowed the dewar to warm up a bit, but not so much that it reached a temperature threshold called the triple point, where nitrogen exists simultaneously as a solid, a liquid, and a gas.

After a period of warming the dewar was recooled, but unbeknownst to the engineers, it chilled down unevenly, getting coldest first at the base of the inner can. Here the solid nitrogen contracted, forming a slight gap between the ice and the can's base.

Nitrogen gas that had not frozen seeped into this gap, where it froze. After a series of warm-ups and cool-downs, several layers of nitrogen ice had built up around the outside base of the inner can. During the warm-up phases, the ice expanded and, like a piston, began pushing against the inner surface of the dewar.

The resulting deformation was slight—a mere 4 millimeters—but that was enough to prevent one of NICMOS's three mirrors from coming to precisely the same focus as the others. Nonetheless engineers were confident that the mirrors would return to focus after a substantial fraction of nitrogen evaporated when NICMOS was on orbit. "Nobody thought that this material had any strength," Thompson recalls. "They thought it was like packed snow. So we didn't think it would have any effect whatsoever." In the meantime care was taken to preserve the flight integrity of the dewar by preventing further deformation. To mitigate the optical alignment issue, a new focusing system was installed to give the camera twice the focusing range it had before.

After it was installed onboard the Hubble Space Telescope in February 1997, engineers began gradually warming the camera's dewar. Unfortunately, they soon found out that the "packed snow" pressing in on the dewar was, in fact, rock hard. The expanding piston of ice further skewered the dewar and defocused the mirrors. The ice also pushed the top of the NICMOS camera baffle against the shield holding the filter wheel, which operates at a warmer temperature, further bollixing things up. Ultimately, the "thermal short," as it is called, cut the lifetime of NICMOS by two years. Its last gasp of cryogen puffed away in January 1999.

The experience forced Thompson and his team to literally rewrite the book on dewars. "Once we found this was going on," he said, "we started to check our references. We kept seeing references back to previous [dewar] handbooks, and we kept tracing those back until we got to a single reference that said, *Dewar, 1890.*" This dates back to the time when Sir James Dewar invented the first vacuum flask. A nineteenth-century error had been handed down to twenty-first-century science.

Salvation came by way of a space shuttle mission in March 2002, when astronauts installed a new cooling system that not only extended the life of the camera but also increased its performance. The revitalized NICMOS would in the future operate at a slightly "warmer" temperature of 77 degrees above absolute zero, but its three mirrors would remain stable and in focus for the lifetime of the Hubble telescope.

So this is yet another reason for Thompson's sanguine state on the day of the public release of the HUDF. With the NICMOS Deep Field in hand, he was justifiably proud, and relieved, when he presented the contributions of the once-troubled NICMOS project that morning.

He is also happy to see, for the first time, the HUDF image itself, which complements in every respect the NICMOS image. Keeping the HUDF data under wraps, even from a researcher as close to it as Thompson (the final NICMOS image was assembled from exposures taken with both the near-infrared camera and the ACS), was in accordance with the Hubble Treasury Program's policy of making data available to everyone at the same time. He doesn't mind, though, as long as he gets a decent crack at the data. "There are over fifteen hundred objects in the [NICMOS] infrared image," he tells reporters. "We have our candidates already marked, but then you have to go through the process, which isn't very glamorous. People think you say 'Eureka!' but you have to go through all the ways you could be wrong and do all the checks, all the statistics that take time to do a good job. So people really won't be writing papers tomorrow. At least, they *shouldn't* be writing papers tomorrow."*

I ask Thompson what he sees in the NICMOS data, but even as the words leave my mouth, it occurs to me that I sound like I'm asking a mystic seer to interpret mists in a crystal ball. His reply is appropriately enigmatic: "We may be seeing past the end of the beginning, when galaxies first started to build up. If we could see past that time, where we don't see as many galaxies, then that would be an indicator that we've finally gotten up to that place."

The end of the beginning. No, Thompson is not alluding to the trippy mantra at the end of the Beatles song "Tomorrow Never Knows." Rather, he's talking about seeing the end of reionization, that roughly 500-million-year-long cosmic buildup during which the universe seethed with an ever-growing population of smoldering stars and galaxies. The "beginning," says Thompson, comes near the end of this epoch, "when the universe began assembling galaxies that are pretty much the same as we see today and the number of galaxies is also about the same as we see today. We should be in that period—not at the *very* beginning when the very first stars formed, but in that process where we're starting to build galaxies." This would mean that somewhere in the NICMOS image are galaxies dating to a time between 500 million and 800 million years after the big bang.

In the interest of making sure we are on the right side of the looking glass, it might be useful to provide a brief reductive review of the first 1 billion years of cosmic history, particularly as the se-

---

*In fact, the very next day, March 10, 2004, a paper by Andrew Bunker, Elizabeth Stanway, Richard Ellis, and Richard McMahon appeared on the astro-ph website maintained by Cornell University reporting their analysis of the star-formation rate in the universe at redshift 6 *based on the new HUDF data*. It was the first HUDF paper published by outside researchers. Bunker says they pulled an "all-nighter" to get it written.

quence of events has been significantly revised by new observations in recent years.

For the universe to be "reionized," it stands to reason that it must have been ionized once before—and it was, 13.7 billion years ago, in the aftermath of the big bang. From a quantum point—so goes the theory—the universe erupted into a rapidly expanding mass of hot gas called plasma. Sometimes referred to as a fourth state of matter, plasma contains nearly equal amounts of free electrons (which are negatively charged) and free protons (which are positively charged) that move about independently of one another. In the violent conditions of the just-born universe, it was impossible for elementary particles to get their bonding act together to form atoms. They were knocked randomly about like frenzied bumper cars. Some 300,000 years later, however, temperatures within the expanding plasma cooled to about 3,000 kelvins (5,000 degrees Fahrenheit), which is roughly half as hot as the surface of the Sun. This temperature was low enough for the protons to finally link up with electrons to form neutral hydrogen, the most abundant element in the universe. All the while, expansion was stretching the light from the primordial fireball ever thinner, like smoke on the wind. As increasing numbers of particles merged to form neutral, and opaque, hydrogen gas, and as expansion and cooling continued unabated, the light of the initial big bang began a 500-million-year fade to black, ultimately plunging the universe into the Dark Ages.

It didn't stay that way for long, though. Between 100 million and 300 million years later, and with universal expansion slowing even more, dark matter began to work its magic by gravitationally collapsing into discrete aggregates called dark matter halos. The size and mass of these halos are highly debated, but not their effect. Acting like cosmic dust bunnies, they exerted enough gravity to overcome expansion at local scales and thus attract neutral hydrogen gas. Depending on the subatomic essence that actually constitutes dark matter (one of the leading exotic candidates is the neutralino, the lightest of the supersymmetric particles), the halos' drawing power could have been prodigious at relatively modest scales. Computer simulations show that dark matter halos with the meager mass of the Earth might do the trick. One can imagine such an agglomeration, a single gaseous thrombus embedded within a greater vascular network of infalling hydrogen gas.

That's when the real magic occurred, for as this and other dark matter halos collapsed, the flash point for the first star was reached. It could have been one star or even a cluster of massive stars—nobody knows for sure. Whatever it was, the growth medium of this *astrum primordium* would have been a rich one— pure neutral hydrogen at great densities. At the same time, though, this kind of growth medium would also tend to encapsulate newborn stars by absorbing their photons. Without enough photon power to pierce their surrounding shells of neutral hydrogen, they might live out their lives entirely sheltered within their gaseous cocoons. The hotter, more massive sources, however, would generate ionization "fronts" that would eventually eat through the hydrogen in their halos, allowing their photons to escape and begin reionizing the universe.

Over the following 200 million years, the Dark Ages would gradually subside. Robust ionization fronts would percolate out from hot stars, saturating the surrounding intergalactic medium with radiation. Expanding ionization fronts would begin overlapping with other ionization fronts, dividing and conquering the universe millions of light-years at a time into ionized bubbles surrounded by still larger shells of neutral hydrogen.

Finally, about 1 billion years after the big bang, nearly every low-density region of the universe would be fully reionized, ring-ing down the curtain on the reionization epoch and ushering in the Cosmic Renaissance. The fog of neutral hydrogen would lift, and the universe would become transparent. The high-density regions that did not contain stars would remain pristine because the atoms in their gas could recombine fast enough to shield against external radiation. These would form the seeds of new galaxies that, 13 billion years later, would surface in the Hubble Ultra Deep Field and NICMOS images.

Reionization, however, may not have been a one-time occurrence. According to some astronomers, and data from the Wilkinson Microwave Anisotropy Probe (WMAP), the universe may have been reionized twice. This idea was first proposed in 2003 by Renyue Cen of Princeton University, and subsequently bolstered with observations by J. Stuart B. Wyithe of the University of Melbourne, and Abraham Loeb of Harvard University, then at the Institute for Advanced Study in Princeton. A two-stage reionization process would be a product of both the density of the neutral hydrogen medium and the life spans of the first stars.

The first reionization phase would have been sparked by a generation of metal-free, short-lived suns called Population III stars that initially formed around a redshift of 40, or some 60 to 70 million years after the big bang. WMAP data indicate, with 68 percent confidence, that the universe was ionized between redshifts 12 and 22, perhaps by these very same stars. But their radiation and eventual explosions as supernovae could have inhibited the development of collapsing clouds of gas that might have later gone on to make more stars. This process, called "negative feedback," has been observed in the Orion Nebula and its environs. (On the other hand, star formation can also be stimulated by these very same processes. When that happens it's called "positive feedback." Astronomers admit that the details of these processes, which work in opposite directions, are not well understood.) At any rate, if negative feedback predominated early on in the universe, there might not have been enough ionizing photons generated to completely ionize the gas. The universe might have experienced a radiant lull during which the hydrogen simply recombined to become neutral and opaque again. A subsequent generation of stars with more sustained lifetimes would have to finish the job by a redshift of 6.

The end of the cosmic Dark Ages is Rodger Thompson's domain, the murky shadowlands where reionization tapers off and galaxies turn on in earnest. It may one day be possible to peer deeply into this twilight zone to find galaxies populated by even earlier stars, but until then the NICMOS image is as close as astronomers can come with a conventional telescope. If the theories are correct and the cosmological models reliable, a deeper survey, an "Ultra-*Ultra* Deep Field," would only find fewer galaxies past Thompson's realm.

But what if, in fact, *more* galaxies are seen beyond this end of the beginning? Thompson spreads his arms in a kind of anything's-possible gesture. "If 300 million years after the big bang we still see galaxies, then we're going to have to say, 'Wow, something just sort of turned on galaxies, like *that*' "—he snaps a finger—"that's not something we expect, but what we expect and what we see could be something very different."

Three months later, during an American Astronomical Society meeting in Denver, Thompson has some news. If their interpretations are correct, the NICMOS team may indeed be seeing galaxies tapering off into the Dark Ages. Of the 1,518 galaxies studied in the NICMOS image, three have colors that indicate a redshift range between 7 and 8. If these galaxies are truly at these redshifts, they are extremely young, about 600 million years old relative to the big bang.

Intriguing though his results may be, they are also a little controversial. The redshift 10 galaxy found in the Southern Hemisphere is not considered by everyone to be an ironclad discovery. The object is extremely faint, and consequently its spectrum is tainted with unwanted signals that astronomers call noise. A noisy spectrum is often unreliable. Such findings must be confirmed first, and only the Hubble telescope can do that—if it survives into the second decade of the twenty-first century.

Thompson's high-redshift galaxies are a little nearer, or older, than the redshift 10 candidate, but when we are talking about the very-high-redshift universe, they don't differ by much, as far as time is concerned. Matter and space flew out of the big bang in all directions in an instant, unfolding and stretching outward or, more precisely, creating an "outward" for matter to expand into. But as it spread from its singular source, gravity began reining in matter and slowing expansion. Hence the time between redshift epochs is compressed the further back you go because the expansion rate was much higher. For example, the difference between redshifts of 0.5 and 0.7, when the expansion rate was more leisurely, corresponds to a little over 1 billion years. The difference between redshifts of 8 and 10, when expansion was clearly more allegro, amounts to just 17 million years; between redshifts of 10 and 20, the time gap shrinks to just 3 million years.

For Thompson, though, the redshift matters less than what is going on in those distant realms. At a redshift of 10, the universe was a thousand times smaller in volume than it is now. "It is sort of a contest," he says. "There may have been fewer galaxies [in the early universe] at that time, so that would decrease the interactions, but the fact that they're closer together means that there would be more chances for interactions. So that's the very thing we want to look for, to see if galaxies at high redshifts are highly dis-

THE REDSHIFT-LOOKBACK-AGE RELATIONSHIP

Redshift	Lookback time (LYs)	Age of universe (years)	
0.001	14 million	13.652 billion	
0.005	70	13.597	
0.01	135	13.529	
0.05	665	13.001	
0.1	1.286 billion	12.380	
0.4	4.256	9.409	
0.5 ♋	5.019	8.647	
Half the age of the universe			
0.7	6.293	7.373	
0.9	7.303	6.362	
1.0	7.731	5.935	
1.2	8.462	5.204	*Redshift Desert*
1.5	9.320	4.346	
2.0	10.324	3.342	
2.5	10.999	2.667	
3.0 ⚹	11.476	2.190	
4.0 ⚹	12.094	1.571	
5.0 ⚹	12.469	1.197	
6.0 ⚹	12.716	950 million	*reionization complete*
7.0 ✸	12.888	778	
8.0	13.014	652	
9.0	13.110	556	
10	13.184	482	
11	13.243	423	
15	13.391	274	
20	13.484	182	*reionization begins?*
30	13.565	101	
40	13.599	66	
50	13.618	47	
75	13.640	25	
100	13.649	16	
1000	13.665	436 thousand	
1200 ♓	13.666	321	*Dark Ages ensue*

♋No significant galaxy evolution until here.

⚹ Star-formation rate is 6 times less vigorous at redshift 6 than at redshift 3.

✸ Bulk of star-forming galaxies that reionized the universe lie in this region.

♓ This approximates the decoupling era after the big bang.

torted, indicating that they have been interacting with each other. And we can see that in the star-formation rate, too, because interacting galaxies produce stars very prodigiously. Right now we think that mass interactions occurred at redshift 2 rather than 10 simply because there are larger galaxies at that time, even though the universe was a lot bigger."

With something as long ago and far away as primordial stars and galaxies, nothing can be taken for granted because they constitute a frozen vista that itself may be part illusion. The early universe no longer exists, except as filmy gradations in time. To see change among the galaxies would take millions of years. If, four hundred years ago, Galileo's telescope could have somehow recorded the galaxies crowding the image of the Hubble Ultra Deep Field, they would have looked exactly the way they do today.

Astronomers face similar challenges in trying to understand the physical processes at play in the early universe. It is unlikely that all stars formed one way or that galaxies were whipped up following a single structural recipe. Images like the HUDF and the NICMOS field certainly provide insights but little in the way of definitive answers. Like most scientists, astronomers are enamored, indeed lured, by the siren's call of simple explanations, but there are few to be found among the shoals of high-redshift galaxies awash at the end of the beginning of the universe. At least as of yet.

"If you are uncomfortable with uncertainty," says Thompson with his Buddha's grin, "then this is the wrong profession to be in."

COSMIC CANVAS

The Darwinian revolution knocked out the back wall, revealing eerie lighted landscapes as far back as we can see. Almost at once, Albert Einstein and astronomers with reflector telescopes and radio telescopes knocked out the other walls and ceiling, leaving us sunlit, exposed and drifting—leaving us puckers, albeit evolving puckers, on the inbound curve of space-time.

—Annie Dillard, "Life on the Rocks," from *Teaching a Stone to Talk*

The light of distant galaxies is often referred to as "fossil light." And that's what it is, in the sense that we are seeing these objects as luminous relics of the past. But the Hubble Deep Field and Ultra Deep Field have pushed back the boundaries of space and time so far that we must now consider fossil light not as old but as young, newborn light. It no longer makes sense to run the film backward from our epoch to the birth of the universe to gain knowledge about how the cosmos sprang into being. We have to begin at the beginning, and run the film forward and hope it all falls into place, leading from there to here.

But let's face it: for most of us, these fossilized galaxies are distant abstracts, beyond imagination and certainly beyond our understanding from what we know about normal galaxies. Here we are, still trying to understand the nearby universe of galaxies, but now we must investigate the nature of objects that are so removed in space and time they have no regional analog to speak of. How can we adequately describe, much less relate to, the visage of the universe that the Hubble Space Telescope captured in those small chinks of sky?

If mere words fail us, we can be thankful that science does not. The deep fields and galaxy surveys all derive from the same grand palette—cosmic evolution. But though the deep fields and imaging surveys have helped erect an evolutionary bridge connecting the near and far universe, the span makes for a dangerous crossing in some places. It's not so much a matter of astronomers' lack of insight or incomplete theories—although those factors do contribute to making the bridge more like a flimsy catwalk in places. Rather, the shaky parts of the bridge are due in large part to an inadequate substructure of solid observations that embrace wider regimes of parameter space. Whereas the Hubble Ultra Deep Field penetrates to incredible distances and lookback times, it is also extremely narrow in width—about nine square arcminutes. Such deep probes are often referred to as pencil-beam surveys. The sky contains 12.7 million times more area than this: 12.7 *million*. There is nothing astronomers would like to see more than similar deep forays across the entire sky, but given the angular scale of the HUDF and the amount of exposure time required, it would take almost a million years of uninterrupted observing to complete such a survey.

On the other hand, wider surveys like the Hubble telescope's Galaxy Evolution from Morphology and Spectral Energy Distributions (or GEMS) map out an area some eighty times larger than the HUDF, on par with the apparent width of the full moon. This provides a larger representative sample of galaxy types, shapes, and sizes, but at the expense of depth. "GEMS maps the universe over the last 10 billion years, out to epochs when it was only 30 percent of its present age," says Shardha Jogee, of the University of Texas at Austin and member of the GEMS team. "GEMS allows us, therefore, to chart the advanced phases of galaxy evolution leading to the present-day Hubble [classification] sequence. In contrast, the HUDF goes further back in time and provides us with a snapshot of the universe when it was only 5 percent of its present age, when the very first protogalaxies were forming and structure was emerging from chaos. It is only by piecing together the information from different complementary surveys like GEMS, GOODS [Great Observatories Origins Deep Survey], and the HUDF that we can hope to solve the grand cosmic puzzle of galaxy evolution."

"What's missing," says NICMOS principal investigator Rodger Thompson, "is a wide-field *deep* survey." This would amount to a Hubble Ultra Deep Field that covers at least a full-moon-size region of the sky at a time. Such an undertaking, however, would itself take years and is not likely to occur anytime soon, says Thompson. "There's no money for it. It's pie in the sky."

Astronomers may feel temporarily hobbled for want of a more attainable piece of the cosmic pie, but they can take heart in the fact that, for the first time in the history of science, the observational riches they have so far gleaned from one tiny corner of the universe enable them to discuss the scope of cosmic evolution to an unprecedented extent. The fact that astronomers can stand before their colleagues today and present findings about galaxies that existed in a universe barely 700 million years old speaks volumes about how much more of the cosmic landscape has been assayed compared with the tracts of the last decade of the last century. In the early 1990s claims about detecting evolving galaxies would have been made tentatively at best; a decade before that they would have qualified as pure speculation—not because of limited thinking but because of limited light-gathering capabilities and limited observations. Today, however, such claims are commonplace.

Theorists in the early half of the twentieth century, who relied on observations to confirm or refute their assumptions and predictions about the cosmic canvas, were also hobbled, but in a different way than observers. For them, the universe was a mathematical construct in which the various terms were like the gain control on a transmitter that could be turned up or down to adjust the output of their models. Sometimes, however, what worked beautifully when the gain was turned one way did not reflect the reality of subsequent observations. In modern parlance some might say theorists weren't thinking outside their cosmological models. In fact, during the first three decades of the twentieth century, what they had done was paint themselves into an irreconcilable cosmic corner.

The theorist making the broadest brushstrokes in those days was Albert Einstein. In his 1915 field equations of general relativity, he proposed the model of a closed, spherical universe that contained matter but was stable—that is, one that wouldn't collapse on itself. To counteract the relentless attraction of gravity, he added an arbitrary term signified by the lowercase Greek letter lambda (λ) that expressed a force that pushed against gravity's pull. This mathematical contrivance became known as the cosmological constant.*

---

*Today astronomers represent the cosmological constant with an uppercase lambda, Λ.

Although the cosmological constant stabilized Einstein's model, which must have pleased him very much, there was no satisfactory physical explanation for a repulsive force, which no doubt annoyed him no end. Even when he introduced it in the general theory of relativity in 1917, Einstein admitted that it was an unfounded, but necessary, mitigating factor. Later, after Edwin Hubble showed that the universe was expanding, Einstein readily dismissed the cosmological constant as being "theoretically unsatisfactory anyway." In fact, some astronomers speculate that what may have really ticked off Einstein was not his so-called blunder but that he failed to predict expansion twelve years before Hubble discovered it.

A second model of the universe, proposed in 1916 by Dutch astronomer Willem de Sitter, sprang from Einstein's field equations. The de Sitter model, too, was spherical and closed, but there was one important difference: his mathematical solution showed that if "test atoms" were inserted into his universe, they would recede, and as they did so, they would appear to emit energy at lower and lower frequencies. This meant that spectral lines from distant atoms would be redshifted. Moreover, they would also accelerate away from one another because of Einstein's λ term. But while de Sitter's model embraced expansion, his equations indicated that the density of matter in the universe was so low as to be mathematically irrelevant. In fact, the immutability of the universe depended on there being no matter present at all. In a theoretical sense, at least, this was the first whiff of dark matter and its cousin in mystery, dark energy.

By the mid-1920s it appeared that cosmology had reached a cognitive impasse. The eminent English astronomer Sir Arthur Stanley Eddington would later drily sum up the dilemma by saying that astronomers were left with a universe of either "motionless matter" or "matterless motion." It was indeed a problem equal to any in cosmology today.

Two theorists working independently and several years apart finally broke the stalemate: Russian mathematician Alexander Friedmann in 1922 and Belgian astrophysicist Abbé Georges Lemaître in 1930. Both men contemplated, for the first time, curved, nonstatic universes in which mass and energy were distributed uniformly at all scales. Friedmann's work, unfortunately, had been published in an obscure German journal and remained unknown until 1930, when Eddington and de Sitter called attention to it.

In the meantime Lemaître, who was also unaware of Friedmann's work until enlightened by Einstein, formulated his own variation on a curved universe in which the recessional velocities of the galaxies observed by Vesto M. Slipher between 1914 and 1925, and confirmed by Hubble in 1929, were the result of cosmic expansion. But Lemaître made a giant leap—backward. In a letter published in 1931 in the journal *Nature*, Lemaître laid the groundwork for his argument, based on the laws of quantum theory and thermodynamics.

According to quantum theory, energy of a constant amount comes in discrete packets called quanta that behave much like particles of matter. But according to the second law of thermodynamics, the amount of disorder or randomness of energy and matter in a closed system is always increasing, from order to chaos. Lemaître reasoned, therefore, that the number of quanta in the future would also increase. "If we go back in the course of time," he wrote, "we must find fewer and fewer quanta, until we find all the energy of the universe packed in a few or even a unique quantum." He added, "If the world has begun with a single quantum, the notions of space and time would altogether fail to have any meaning at the beginning; they would only begin to have a sensible meaning when the original quantum had been divided into a sufficient number of quanta. If this suggestion is correct, the beginning of the world happened a little before the beginning of space and time." Lemaître speculated that the atomic weight of this unique primeval atom would amount to "the total mass of the universe."

Lemaître's model of a unique quantum from which the universe expanded outward in all directions resonated nicely with another big idea of Einstein's and the relativists of the times, called the cosmological principle.* Einstein had originally applied the cosmological principle as a way of reducing the universe to its simplest terms. This, in turn, allowed him to apply his field equations of gravity and general relativity to deduce the nature and structure of the cosmos, at least on paper. Philosophically, the cosmological principle also constituted the final banishment of mankind from the geocentric garden by painting the universe onto a canvas spanning the largest of scales, where our galaxy was reduced to a subvisual part of a homogeneous whole.

The cosmological principle states that the universe looks basically the same in all directions from any location. This means that hypothetical observers located in other galaxies see a universe that on the whole looks more or less the way it looks from our galaxy. Except for local variations on scales smaller than galaxy clusters, there are no preferred locations. The properties of the universe are the same everywhere, or as astrophysicists would put it, the universe is homogeneous (looks the same to every observer) and isotropic (looks the same in every direction). The cosmological

*The cosmological principle was originally introduced in 1932 by English physicist Edward Arthur Milne in a theory he called "kinematic relativity." Although Milne's formulation, a variation of general relativity, never caught on, his notions about how the universe would look to other observers from other galaxies certainly did.

principle also applies to an expanding universe in that all observational frames of reference change uniformly with expansion. Thus all observers everywhere would see the same relative numbers of galaxies receding from their locations at similar expansion rates.

By 1932, with cosmic expansion now factored in, cosmologists had conflated the best of de Sitter's and Einstein's ideas into a single version, the "Einstein–de Sitter" model. In this incarnation, the universe is flat (the curvature is neither positive nor negative), expands continually (but at an ever-decreasing rate), and has no cosmological constant. For decades the Einstein–de Sitter model prevailed as the "standard paradigm" (a fond phrase often used by theorists). For the next sixty-six years all was right with the universe. Then things took a strange turn in 1998 with the discovery that—guess what?—the expansion of the universe wasn't slowing down at all but appeared to be accelerating, pumped up like some kind of cosmological cyclotron. Shades of Einstein! The astronomical community realized their models would have to be revised and some form of the cosmological constant reinstated.

After all the awards, backslapping, and accolades were conferred upon those who had made this exceptional discovery, astronomers settled down to grapple with the new reality. At least the laws of physics still held true. There would be no academic equivalent of a Wall Street crash, with astrophysicists throwing themselves off the catwalks of observatories. The universe could still be considered as flat as Kansas—flatter, even. And although the mass density of the "new" universe was far below that of the Einstein-de Sitter model, it was undeniably closer to de Sitter's original archetype. This might not be such a bad discovery after all.

The irony doesn't escape Princeton cosmologist P.J.E. Peebles. In a 2005 essay he writes, "The evidence that the mass density is lower than in [the Einstein–de Sitter model] would not have surprised de Sitter. Would the evidence that the low density is compensated by [a cosmological constant] rather than space curvature have disturbed or pleased Einstein?"

We will, of course, never know, though we can speculate that perhaps Einstein wouldn't have been surprised. At the time, he and other theorists were hoisting their cosmological models onto complex frameworks of mathematics. The resulting structure could be radically altered by a single term in the equations. Anything must have seemed possible, or impossible.

The cosmological principle began gaining empirical credence in the wake of a galaxy survey that, appropriately enough, Edwin Hubble himself conducted in the 1930s. You might say that his Herculean survey was father to all deep fields that followed. Using the 100-inch Hooker Telescope at Mount Wilson Observatory, Hubble made deep photographic exposures of more than twelve hundred sample regions of sky and methodically compared the results. He discovered that the number of galaxies was not only similar in each sample but also increased in number with intrinsic faintness. Hubble concluded that, viewed globally, the universe looked more or less the same in all directions and, furthermore, would look the same to an observer in a galaxy 1 billion or 2 billion light-years away. If this were not the case, then we'd see galaxies crowded into a few locations in space and conspicuously absent from others. But we don't see that. Rather, as Hubble wrote in his 1936 book *Realm of the Nebulae*, "We now observe . . . a vast sphere, through which comparable stellar systems are uniformly distributed. There is no evidence of a thinning-out, no trace of a physical boundary."

Further support for the cosmological principle came with the first Hubble Deep Field in 1996. Galaxies seemed to be strewn like windblown eiderdown throughout the field. Still, that fact alone could not validate the cosmological principle. After all, the Deep Field covers a smidgen of sky and shows how the universe looks in only one direction over a small, albeit deep, region of observable space. Astronomers wondered whether a deep field in the *opposite* direction would appear comparable. If so, that would strongly affirm the cosmological principle; if not, the idea that we live in a homogeneous, isotropic universe might have to be scrapped.

In October 1998 the Hubble Space Telescope made a deep field exposure of equal duration and angular size in the southern sky. The region selected this time was in the constellation Tucana, near the south celestial pole. Save for a single quasar, it, too, at first appeared devoid of known galaxies. The resulting exposure was a stunning counterpart to the northern deep field. Both images contain about the same number of odd-looking galaxies, although the Hubble Deep Field South holds a larger population of older and larger galaxies. Despite the disproportion in galaxy type, however, most astronomers agree that the north and south deep fields, taken together, uphold both the cosmological principle and the idea that galaxies have evolved since Lemaître's primeval atom set the universe in motion.

This conceptual conjunction between the cosmological principle and the big bang does have its limitations, though. Strictly speaking, the big bang theory was initially postulated to explain expansion, not the formation or evolution of the universe. As such, in its original form it does not qualify as a "cosmology" as the word is generally defined; that is, it doesn't describe the origin, history, and large-scale structure of the universe. Rather, it is invoked as a kind of deus ex machina to explain how galaxies were endowed with recessional motions. Its evolution into a convincing approximation of the real universe, says P.J.E. Peebles, has come about only recently with more observational evidence.

"Many of the new tests require the postulate of the cold dark matter model," he says. "Still more evidence is needed to turn these additional postulates into parts of a well-established reality."

Whatever reality you subscribe to, neither the big bang nor the cosmological principle explains the existence of galaxies. As de Sitter pointed out, even a universe of dissociated particles or one that was entirely absent of matter could qualify as isotropic and homogeneous. But both concepts surely must be related. As such, they can be used to explain not only why the universe looks so unvaried in all directions, but why matter chooses to be lumpy and disproportionate at small scales but not at the largest of scales.

In its simplest form, the big bang theory describes a seminal, eruptive event in which the universe abruptly sprang forth from a region of space smaller than the period at the end of this sentence. How do cosmologists know this? The answer is they don't, but they can see the galaxies all rushing outward over time, and as with old Lemaître, it's no mental stretch to see how the expansion movie would play out if it were reversed. Eventually all the matter and energy making up the present universe would have to occupy the same small volume. All combined, it would be unimaginably massive, and gravity, being the dominant force, would overcome radiant energy and compress the universe into an infinitely dense point, like the singularity at the heart of a black hole. Except that *this* black hole, for some reason, was thrown violently into reverse.

After making its bombastic entrance, the universe settled down to a languid, insipid period of expansion with no other purpose, apparently, but to distend, which it has done for at least 13.7 billion years.

Although astronomers cannot look back and "see" the big bang itself, they can do the next best thing: they can study what remains of its radiation, which is best detected in the microwave part of the spectrum. Called the cosmic microwave background, or CMB, this is the grandest, most expansive cosmic canvas of them all. No matter where you point your radio or infrared telescope, this primordial glow, at large scales, is extraordinarily uniform in every respect. There are neither large temperature deviations to give it radiant diversity nor salient wrinkles to give it topographical relief. In fact, one of the most important cosmological predictions the big bang theory makes is that the CMB will be practically featureless. And so it is. Before galaxies, stars, and planets came along, the universe was as smooth as a billiard ball.

The cosmic microwave background was predicted in the 1940s and finally discovered, accidentally, in 1965, but its temperature wasn't precisely nailed down until the Cosmic Background Explorer, or COBE, measured it from space in 1990. An instrument aboard COBE, called the differential microwave radiometer, essentially took the temperature of the universe and found that it hovered slightly above absolute zero—2.72 degrees kelvin, which is some 100 million times cooler than room temperature. Moreover, the temperature was uniform across the sky to better than one part in a thousand.

The smoothness and all-sky predominance of the CMB provide the best evidence yet that the universe evolved from a hot primordial state. But that isn't all these qualities signify. Astronomers can also look at the radiation as the first single structure to appear in the pregalactic universe and map its features as one might map the surface of a new planet. Etched into the CMB is a fine-scale matrix of overdense and underdense regions—density fluctuations—which manifest themselves as very slight temperature variations above and below that of the 2.72-kelvin mean. The overdense regions act like a gravitational well for the background radiation's photons; thus they appear slightly cooler than the underdense regions. The temperature variations are extremely trivial—two regions of the sky might differ by 0.0002 of a degree or less—but they nonetheless amount to the CMB's hills and valleys.

Astronomers extracted their most detailed view of these density fluctuations to date in 2003 using the Wilkinson Microwave Anisotropy Probe. Launched in 2001, WMAP's instruments are forty-five times more sensitive than COBE's. The probe scanned the sky for twelve months using a pair of telescopes tuned to five discrete microwave frequency bands between 22 and 90 gigahertz. The data were then painstakingly compiled into an all-sky football-shaped map called a Mollweide projection. The result was a Jackson Pollock–like spatter portrait of the universe when it was only 380,000 years old. WMAP's ability to measure temperature fluctuations at much higher angular resolutions brought into sharp focus primordial structures that were only blurry patches in the COBE maps. These squiggles, said enthused astronomers, amounted to the seeds of what would later grow into the vast structure of galaxy clusters seen throughout the universe. The late John Bahcall, an astronomer with the Institute for Advanced Study in Princeton, said the achievement was comparable to measuring a fifty-year-old man as a baby only twelve hours old, "right down to the total body weight, the length of the legs, the size of the ears, and the amount of hair."

In light of these impressive findings, it appears that the cosmological principle, which rose to prominence three-quarters of a century ago, is holding up very well. The WMAP data show that if you take a box billions of light-years to a side and measure the average temperature and density of matter in the box, you will derive the same readings wherever the box is located in the universe. On the largest scales, at least, the universe is uniform and unvarying.

But the WMAP data cuts two ways. Look deeply enough at any apparently smooth thing and at some point you're bound to encounter structure of some kind. From space, for example, the ocean looks calm, but from the deck of a small boat it's all wave, foam, and nausea. The same is true for the universe. On the whole, WMAP took a picture of the cosmos that shows it to be essentially as smooth as glass. But turn up the contrast enough to reveal those trifling over- and under-dense regions and it becomes immediately apparent that the glass is etched with structure. A *lot* of structure. How did it get there? Something had to disturb the otherwise tranquil surface of the early universe so that matter could begin to clump into stars and galaxies. The question is, what?

Some theorists say there were inherent defects in the early universe that could later pull matter together, perhaps into finite lines of fundamental particles. This is the basis of the speculative "string theory." Proponents say general relativity works only if we ignore quantum effects and describe the universe in terms of purely classical physics. String theory, they say, incorporates both quantum mechanics and general relativity.

Others favor another solution that is just as bizarre but has more supporters. It's a kind of hyperexpansion expedient to the big bang called inflation.

Inflationary models invoke some heavy-duty physics, like gauge symmetries, spacetime defects, and phase transitions, the gory details of which need not concern us here. Since its inception in the 1980s by MIT physicist Alan Guth, inflation has been through a number of variations, but the essence of the theory remains this: less than a wink after the big bang occurred, the universe underwent an instant of flash inflation in which its volume leaped from atomic to cosmic scales. The presto, change-o! moment occurred 10^{-35} second after the big bang (or 100 millionth of a billionth of a billionth of a billionth of a second), when the universe had "cooled" to a temperature of 10^{27} kelvins. This thermal decrement allowed three of the four fundamental forces of nature—the strong nuclear force and the weak and electromagnetic forces (collectively called the electroweak force)—to go their separate ways. When that happened, the dissolution released an enormous pulse of latent energy, lasting all of a thousandth of a billionth of a second. The result was a big whoosh, which enlarged the universe by 10^{75} times its previous size, some 10^{50} times more voluminous than that predicted by the big bang theory alone.

The beauty of inserting an inflationary moment in the early universe is that it resolves some of those just-right issues raised by an uninflated version of the big bang. Before inflation, all particles in all regions were so tightly packed together that the distances between them could have been traversed at the speed of light and therefore they must have been in causal contact. After inflation, though, these regions were abruptly conveyed to voluminous distances well beyond their mutual causal horizons. This disconnection of previously connected regions explains why the universe appears flat and Euclidean and why the galaxies seen in the deep surveys in opposite directions in general exhibit similar properties, even though they can't possibly have interacted with each other in the distant past.

Inflation also resolves why the universe at large scales looks creamy smooth, despite the fact that lumpy galaxies exist at smaller scales: the matter-energy distribution was, in fact, not completely smooth but interlaced with slight density variations at *quantum* scales, too small to effect changes at larger scales. After inflation, however, these quantum fluctuations were greatly amplified. Inflation theory asserts that the largest fluctuations had to have angular sizes of one degree in order to correspond to the scale of structure observed today. The precise WMAP results verify this prediction. Hence what started out as infinitesimal foamy waves of particles merged with other waves after inflation and mounted into rolling swells of galaxy clusters and filaments seen today at very high redshifts.

Unfortunately, as physicists point out, the Achilles' heel of inflation is its lack of direct evidence; none exists. That doesn't mean, however, that prospects for that evidence don't exist. Gary Hinshaw, a member of the WMAP science team, thinks some progress may be made on that front.

"If inflation occurred at a suitably high temperature or early time," he says, "it should have left behind a background of ripples in spacetime in the form of very low-frequency gravity waves. These waves would produce a distinct, but faint, imprint in the polarization of the CMB that might be observable by a future satellite mission."

The mission to which he refers is called the Beyond Einstein Inflation Probe, which is currently plotted on NASA's ten-year road map, although it may be derailed by the Moon-to-Mars Initiative proposed by the Bush administration. If it isn't scrapped and is ultimately successful in detecting this gravitational imprint, it could provide very strong evidence for inflation. "If it is not observed," Hinshaw says, "it means that inflation occurred at a lower temperature—later in time—or not at all. Both prospects are very exciting."

Another objection held by some physicists is that inflation could work only if space were largely homogenous *before* inflation occurred. Homogeneity, then, would be a prerequisite, not an effect, of inflation. And though inflation can explain the slight defects in the CMB as quantum fluctuations, those, too, must be considered a precondition. If inflation did occur, it is guilty of being fine-tuned.

"It appears to be true that inflation must be fine-tuned in the sense that the 'inflationary potential' has to have a certain form to reproduce some of the details of the CMB," says Hinshaw. "From this point of view, I think one can only hope that inflation—or something like it—emerges as a natural consequence of string theory—or something like it—and that the form of the potential is mandated by the underlying physics."

Not having all the answers is never as bad as not having the means to answer the questions. New and powerful telescopes on the ground and in space are already making important headway toward correlating the oldest light in the universe with the origin of stars and galaxies. This is not a recent objective, though the targeted goals are new. Since the days of Galileo, astronomers have been scouring the cosmic canvas for answers to great questions. This is the first time in that long history that technology has finally caught up with their aspirations. And when it comes to the universe, timing is everything.

THE EGG BEFORE
THE CHICKEN

In order to understand what are the earliest building blocks of galaxies like our own, one must detect and identify "first light" sources,
i.e., emission from the first objects in the Universe to undergo star
formation.

—Nino Panagia, "Detecting Primordial Stars"

One of the ironies of the current state of cosmology is that more is
known about the universe when it was 300,000 years old than
when it was 1 billion years old. The reason: an entire epoch in the
early universe has been overlooked, albeit not purposely. The most
distant galaxies and quasars so far detected exist at a time when the
universe was a few hundred million years old, or 3 or 4 percent of
its present age. Microwave and radio telescopes, as we have seen,
allow cosmologists to look back to a time when the universe was a
little less than 0.003 percent its present age—about 380,000 years
old—when the last photons generated in the big bang fireball were
ricocheting off atomic particles in the cosmic microwave background. The very skin of this alluvial structure, called the last scattering surface, was first measured in detail in 2002 by a radio

telescope at the south pole called the Degree Angular Scale Interferometer, or DASI. The last scattering surface is the big bang's final bow, the moment when the microwave photons broke free from an effluvium of electrons and protons created in the fireball. It even has a redshift: a whopping 1,100.

Unfortunately, the subsequent 1-billion-year history of the universe—the realm lying between the last scattering surface of the CMB and the outermost known galaxy—is literally shrouded in darkness by a pall of neutral hydrogen that obscures the light of any stars or galaxies that may be present. Hopefully this will change sometime after the scheduled launch of the James Webb Space Telescope in 2013. This effort, combined with a new generation of large ground-based telescopes, should provide cosmologists with direct views of objects abiding in this murky realm. It is here that the first light sources detectable—though not necessarily the *very* first to exist—may be found.

What manner of objects will they be: stars, hypernovae, protogalaxies, mini-quasars, black holes, or something entirely unknown? Observations today offer tantalizing, though sketchy, possibilities. For now, the rest of this cosmic no-man's-land must be fleshed out by theory. Some astronomers are confident that mammoth, volatile stars on the order of hundreds of solar masses formed less than 200 million years after the big bang. Sometime later the first supernova explosions, a hundred times greater than conventional Type II supernovae, rocked the dark, primeval universe. These stellar eruptions would have been the architects, builders, and raw material reserves for all the stars and galaxies that followed.

The impact of such heavyweight stars on the rest of the universe would have been dramatic. They would have radiated mainly in the ultraviolet, which in turn would have changed the physical state of the gas by ionizing neutral hydrogen and, if hot enough, the helium gas in the immediate surroundings. Ionization "bubbles" (astrophysicists call them Strömgren spheres) would have developed, expanding rapidly outward like a wildfire rushing through a dry, dense forest. Over the ensuing hundreds of millions of years, as greater numbers of stars were born, more ionization bubbles would have developed and spread across the universe, meeting and merging with other bubbles until all of the intergalactic gas was ionized. The newfound radiance would have burned off the hydrogen fog, ending the Dark Ages, and the universe would have been ablaze with stars, signaling the beginning of the Cosmic Renaissance.

Backing up these grand theories are a few sound observations. Researchers from Caltech and the Sloan Digital Sky Survey have produced the most supportive results. In late 2001 they discovered that the ultraviolet light of quasars existing at a time when the universe was about 900 million years old was strongly absorbed by neutral hydrogen. This would have been near the end of the Dark Ages, suggesting that observers were witnessing the tail end of ionization and the beginning of the Cosmic Renaissance.

Further corroboration comes from observations made with the Very Large Array radio telescope near Socorro, New Mexico, and Europe's Plateau de Bure radio interferometer, located in the southern French Alps. A team of international researchers discovered that a galaxy existing only 870 million years after the big bang was awash in carbon monoxide gas—as much, in fact, as some galaxies that are relatively nearby. The excess carbon monoxide could only have been exhaled from the cores of an earlier generation of stars that formed about 650 million years after the big bang. "Our discovery of this much carbon monoxide gas in such an extremely distant and young galaxy is surprising," says team member

Chris Carilli of the National Radio Astronomy Observatory. "How do you process so much material through stars and back to the molecular interstellar medium phase in such a short time? Then there is the general issue that the high-redshift carbon monoxide sources all have molecular gas masses one to two orders of magnitude larger than a large spiral galaxy, such as the Milky Way. Are we seeing the formation of a large spheroidal galaxy in a massive molecular-enriched starburst of extreme proportion?"

According to the team's calculations, a population of 10 million metal-free stars each with the mass of the Sun could enrich a galaxy with carbon and oxygen in less than 10 million years. The team considers this a lower limit for enrichment, however, as it takes 100 million years for the gas to redistribute and cool. Following this process back in time indicates that the enrichment process probably was under way by a redshift of 8.

Supernovae, too, could have enriched their host galaxies given a vigorous star-formation rate. Stars with masses hundreds of times that of the Sun live only a few million years before exploding as supernovae. If we could witness this early phase of the universe at a speed of, say, 1 million years per second, we would see supernovae popping off like paparazzi flashbulbs. A barrage of supernovae disgorging processed metals this quickly into the interstellar medium would explain the rapid augmentation of cooked-up elements observed in early galaxies, as well as solve the decades-old mystery of why intergalactic gas is more metal-rich than the galaxies themselves.

The very deaths of supermassive stars could also have contributed to building up early structure. Having exhausted its internal source of nuclear fuel, and hence being unable to support its own weight, a massive star's gaseous envelope collapses onto its core, the violent rebound of which blows the star to atoms. The initial collapse, however, compresses a lot of the star's mass into a volume only a few kilometers across, creating a black hole. If, as computer simulations suggest, the first stars lived and died in the densest parts of the universe, then the proliferation of black hole corpses would act as gravitational snares for other proximal black holes, thereby creating more massive attractors, some with masses millions of times that of the Sun. These, in turn, would sweep stars into swarms, building small protogalaxies and powering up mini-quasars. Over the next half-billion years, collisions between these objects would create still larger and more disturbed-looking galaxies, not unlike those seen in the deep field images.

It is a neat little scenario, but one should beware of pat answers in small packages. No one agrees more than NICMOS principal investigator Rodger Thompson. "I think that astronomers would like to have things very simple," he says. "That is, if you form stars, you only form them one way; if you form galaxies, you only form them one way. But I think Nature's a lot more complicated than that. Nature has many ways of forming things." Thompson, of course, is first and foremost an observational astronomer, and his appreciation for a more complex beginning to the universe is perhaps a by-product of the rich array of young, high-redshift objects seen by NICMOS. Theorists, on the other hand, can afford to focus, even obsess, over one or two scenarios to the exclusion of all others. This is something I realized when I sat down to interview three stellar theorists whose work is to try and determine exactly what kind of stars formed first in the universe.

Over sandwiches and soda I spoke with Aparna Venkatesan and Michael Shull, both of the University of Colorado, and Jason Tumlinson, of the University of Chicago. The trio had collaborated on a 2004 paper published in the *Astrophysical Journal* on the reionizing capabilities of the first stars. Venkatesan, born in Chennai, India, considers herself a theoretical cosmologist, although these days she dabbles in observational work that tests her theories about the formation of the first stars. Her face is graced with a warm, friendly smile but also confidence. (She is, after all, a student of tai chi.) She works with the affable Shull, whose interest in the reionization of the intergalactic medium nicely meshes with Venkatesan's theoretical leanings. Tumlinson, a stellar theorist and avid baseball fan, looks mighty young to be a theorist of any kind, despite the beard, but then again, my perceptions may be slightly skewed by the fact that I see my own youth gaining in redshift these days.

"Three theorists and four opinions," quips Tumlinson as I roll the tape recorder. But that's not exactly the case. Tumlinson and his colleagues are of a mind that the first stars to form in the universe didn't have to be as massive as some astrophysicists have proposed. That conclusion was largely drawn from the first round of WMAP results, which had suggested that the universe was reionized perhaps as early as redshift 20. But for this to occur, the first stars would have to have been colossal, perhaps as much as a thousand times the mass of the Sun. "I think that was sort of a theorist's prejudice in that these things are very appealing theoretically and are justified by the simulations of early galaxies," says Tumlinson.

But—and this is the point Venkatesan, Tumlinson, and Shull want to make—the first stars wouldn't necessarily have had to be very massive. Instead, the researchers think that a population of stars between 10 and 30 solar masses could have reionized the universe in accordance with the WMAP findings as well as studies of the spectra of distant galaxies, which indicate reionization occurred around a redshift of 6 or 7. If true, this would also agree with the GRAPES findings, mentioned in Chapter 2, that the youngest stars would not have had to be very massive but instead could have been similar to stars in the modern universe.

On the other hand, the first stars *would* have to have had significantly different chemical compositions. They would have been pristine objects born of pristine material, and as such they would have originated in an era before any stars had forged in their cores elements heavier than helium. The absolute lack of such "metals" would be the hallmark of the very first stars.

All stars contain some traces of metals, some more than others. Compared to the Sun, for example, which is considered the fiducial metal standard, high-mass stars that have greedily and rapidly reprocessed material forged by earlier stellar generations are metal-rich. Astronomers call these Population I stars. They typically appear in the arms of spiral galaxies, where star formation is galvanized by copious amounts of interstellar gas and dust. On the other hand, the long-lived red dwarf stars that live high in the remote and isolated recesses of our galactic halo have metallicities that are between 0.001 and 0.00002 that of the Sun's iron abundance. These are called Population II stars, and they are truly ancient; some are nearly as old as the universe itself.

Population II stars have scant metallicity values, to be sure, but the first stars would have had far less; in fact, they would have had metallicities of 0. This would have made them very different beasts from the youngest stars visible even in the myriad galaxies of the Hubble Ultra Deep Field. Consequently, astronomers call them— that's right—Population III stars.

Essentially, no matter which population group a star belongs to, it is essentially a huge ball of gas in a delicate push-pull balance. On the one hand gravity wants to collapse the star, while on the other radiation wants to inflate it like a balloon. Without these two opposing forces, a star couldn't exist. It is the mechanism that generates the radiation, together with a star's mass, that determines the star's size, luminosity, evolution, and fate. The internal energy source arises from two nuclear processes: the collision of two hydrogen nuclei (protons) building up into helium, called the proton-proton chain; and a series of reactions catalyzed by carbon, called the carbon cycle, or the carbon-nitrogen-oxygen cycle because the cycle involves nitrogen and oxygen in the transition.

In stars with the mass of the Sun (which, in case you were wondering, is considered to be an older Population I star), the proton-proton chain predominates. But because the carbon cycle is very temperature-dependent, it is the prevailing energy source in massive Population I stars with core temperatures above 20 million kelvins.

Population III stars were not only much hotter than this, they were also forty to fifty times more massive than the Sun. Therefore one might think Population III stars, too, would rely on the carbon cycle as their energy reactor. Not so. Remember, when the universe first exploded onto the scene, it was three parts hydrogen to one part helium, and that was it. All of the other elements that make up the periodic table, including carbon, came later, after they were fused in the thermonuclear hearts of stars and spewed into interstellar space upon their explosive deaths. Since carbon did not yet exist, massive Population III stars were forced to resort to the less-efficient proton-proton chain.

But there is yet another hitch. Protons have the same positive charge, like the identical poles of two magnets; thus they have a natural aversion to each other. The only way Population III stars could overcome this resistance was to collapse even further, packing more mass into a confined volume. Now gravity could overrule the protons' antipathy for each other and force them to combine.

The copious amounts of energy consequently released enabled Population III stars to achieve the necessary temperatures to counteract the gravitational infall of their gas. This made them smaller, but much hotter, says Tumlinson.

"If you imagine two stars of ten solar masses, one of which has a solar metallicity and the other metal-free and completely devoid of carbon, the metal-free star will be on average five times smaller and twice as hot on the surface," he says. "Stars basically emit as black bodies, so a higher temperature also translates to a bluer spectrum. So the blue colors, the small sizes, and basically every other cosmological and observational signature of the first stars is a direct result of the fuel source, which is in turn a direct result of their very unusual chemical composition."

This is also why the youngest galaxies in the early universe exhibit a characteristic blue color: they contain populations of hot, low-metal stars. If you could take a spectrum of such a star, it would exhibit a strong Lyman-alpha emission line, but because it produces 100,000 times more ionizing photons from helium, it would feature an even stronger ionized helium (He II) line.

Venkatesan points out that although the first stars may never be found—they lived fast and died young—their cosmic precipitates still exist in abundance. "The conditions right around when the elements were being made in the big bang are being reproduced again in our backyard in the interiors of stars," she says. "They're not that far away. They're in the heart of the Sun and the stars."

Tumlinson draws an elegant balance between the universe and the first stars. "All of the cosmological effects that we've been talking about," he says, "come directly from that one peculiar feature of the first stars. Had the big bang created even a tiny amount of carbon, even *one part per billion*, none of this would be the case.

It's the sort of thing that if you were paying close attention in your stellar structure course, you got it. But it took a while for us to get our brains around it. It's a bit counterintuitive."

In short, these highly radiant 10-to-30-solar-mass metal-free stars, placed in a universe that is far smaller and denser than the one today, would be enough to reionize more than half the universe by a redshift of 30. That jibes well with the first round of WMAP findings, which places complete reionization somewhere between a low redshift of 12 and a high redshift of 22.

Population III stars wouldn't, however, comprise *the* first objects. As Shull puts it, "There had to be an egg before the chicken." Long before a redshift of 30, around a redshift of 50 or even earlier, clumpy dark matter halos—consisting of weakly interacting massive particles (WIMPs), light neutrinos, or some other elemental exotica—would have collapsed under their own gravitational attraction. With masses on the order of 100,000 to 1 million times the Sun, they would have acted like gravitational sinks for baryons, which is the name given to the collective garden-variety protons and neutrons that make up normal matter.

"The baryons trace the dark matter halos," says Venkatesan. "We know that at 300,000 years they had not been clustered to a higher degree than 1 part in 100,000. So even the dark matter halos normally advance to one part in a hundred thousand from homogeneity. But that's all the start they need."

Says Shull, "Things really start popping off at a redshift of 30 and build exponentially to a redshift of 10." A half-billion years later the metal-free party is drawing swiftly to a close. "Redshifts of 4 or 5 might be too late for metal-free stars to continue forming."

Tumlinson says the duration of the metal-free heyday is extremely difficult to determine because it involves so many poorly understood processes—namely, how metals are distributed by supernovae, how metal-enriched gas is mixed with pristine gas, and the details of how galaxies evolve and merge.

"In the study we presented at the [June 2004] American Astronomical Society meeting, we included some simple calculations Mike did to estimate the time it would take for the first stars to enrich their own small protogalaxies . . . and how long it would take their metals, if blown out by supernovae, to reach nearby pristine halos by traversing the intergalactic medium. The former process can curtail metal-free star formation in roughly 10 million years; the latter process takes about 100 million years. These numbers mean that if metal-free star formation starts in earnest by a redshift of 20, it could be over before a redshift of 10."

If the enrichment of other halos by the first stars takes as long as 100 million years, Shull speculates, it's even possible that some areas of the universe have escaped being enriched. "If two regions are racing apart with the Hubble expansion," he says, "then it's even harder to move [metals] from one object to another. That's why I hold out some hope that there is a little pocket of the universe somewhere that's still pristine. Some little dwarf galaxy that hasn't turned on yet that was racing away from its neighbors and was left unpolluted."

Shull almost sounds wistful as he says this, probably because he knows that the odds of finding such objects are not good. If such unsullied hollows exist, they would be extremely small—probably about the size of a typical globular cluster—and very rare. The place to search for them, he suggests, would be in the voids, the vast vacuous regions in the spongy medium of intergalactic space where matter never clumped up. You'd look for the "real loners," he says, where dark matter halos have aggregated a little bit of gas but have not yet turned on. A good radio survey of the nearby universe at the wavelength most sensitive to neutral hydrogen emission—21 centimeters—might be capable of detecting these primordial, pristine wisps of gas.

This is not a far-fetched notion. First, as mentioned earlier, stars of extremely low metallicities—several thousandths that of the Sun—exist today in the outback of our galaxy's halo. Second, and directly related to the first, these stars are extremely old. Some astronomers think that they may have been enriched by only one previous stellar population, or even one star. Hence they could be direct descendants, the immediate offspring, of the first generation of stars to form in the universe. What little carbon or nitrogen they possess would be thanks to the stars that smoldered in the dusky waning years of the Dark Ages.

Another glimmer of hope that some dark, pristine solitaires still exist is the fact that our own galaxy has too few satellites. If you hold that big galaxies should form by merging with smaller galaxies over time, as described in the hierarchical model of galaxy formation, you can estimate how many satellites a galaxy with the age and mass of the Milky Way should have. Turns out we should have a lot more than we do.

"The Milky Way should really have about a thousand dwarf satellite galaxies, and we have nowhere near that," says Venkatesan. "We only have about thirty, and two big ones. So maybe there was a wave of star formation that blew out all their gas and couldn't make any more, or maybe they never started making stars, and in that case they would be these pristine environments. We think they should be out there, but they're just dark."

Two recent discoveries may further bolster speculation about the existence of pristine dark halos or at least of quiet intergalactic backwaters. At the same AAS meeting where Venkatesan, Tumlinson, and Shull presented their findings, a group of Texas astronomers led by Pamela Marcum of Texas Christian University in Fort Worth announced that they had detected two elliptical galaxies that may be of primordial origin. Most elliptical galaxies in the universe belong to galaxy groups or clusters. The ones Marcum's team found, however, exist in isolated environments, with their nearest neighbor being at least 8 million light-years away. Further, the galaxies show no evidence of previous mergers, such as tidal tails caused when one galaxy gravitationally pulls material out of another, or blue colors resulting from merger-induced star formation. This implies that, unless the scars of foregoing interactions have since faded, the two galaxies could have been born as ellipticals.

Just as intriguing are the findings of astronomer Liese van Zee of Indiana University. Using the Very Large Array radio telescope outside of Socorro, New Mexico, she discovered an unassuming dwarf irregular galaxy just 16 million light-years from the Milky Way wallowing in a huge pool of hydrogen gas that, apparently, is uninvolved with the galaxy's prodigious star-formation process. Visually the galaxy, known as UGC 5288, is a patchy bar-shaped system only 4,000 by 6,000 light-years. But at a radio wavelength of 21 centimeters, an emission known as the neutral hydrogen line, the hydrogen disk surrounding the galaxy can be mapped to its fullest extent, which is 28,000 by 41,000 light-years.

There are three possibilities for the origin of this hydrogen nimbus. It could be gas expelled by the galaxy's initial burst of star formation, tidal debris left over from a recent encounter with another galaxy, or the remnant of the material from which the galaxy collapsed. The latter explanation seems the most likely, says van Zee, because "the gas disk is rotating quite peacefully around the galaxy. In terms of frequency, extended gas disks appear to be more commonly associated with galaxies that are unusually compact in their stellar distributions, but even this trend is not universal." If the gas is primordial, unadulterated material left over after the galaxy coalesced, then UGC 5288 could become a target of opportunity for astronomers wanting to study pristine material in the universe and how star-forming gas and star formation are related. It would also suggest that regions of intergalactic space can remain more or less unadulterated over the lifetime of the universe.

"We know that such disks are rare, since only a handful of galaxies are known [to have them]," says van Zee. "I suspect that we will find more as we survey the sky in the neutral hydrogen line."

The detection of the first stars to form in the universe may have to wait until the James Webb Space Telescope is launched in 2013. JWST will have the spatial resolution to actually detect small galaxies populated by these stars, if they can be detected at all. Astronomers may initially see not the rich panoply of galaxies revealed by the Hubble Ultra Deep Field, but unimpressive, isolated reddish blobs. And those would be the nearest ones. Those more distant will likely be dustier, smaller, and consequently fainter, eventually fading from view as observers probe into greater depths of the neutral hydrogen of the Dark Ages.

Such prospects are nonetheless encouraging to cosmological theorists and observers alike. Finding more galaxies in the distant universe would only make things more interesting. But now we're skirting an even bigger issue that was hinted at in this chapter and the one previous: how did all these stars and galaxies form so quickly? Whereas many astronomers are confident that dark matter can facilitate the rapid construction of stars and galaxies, Michael Shull wants to know more about the devil in the details. "How can you form massive quasars, galaxies, or black holes in a few hundred million years?" he wonders. "The really massive stars only live a few million years. So how do you make a billion-solar-mass black hole in the nucleus of a galaxy in only, let's say, 100 million years?"

Just a few years ago this issue might have prompted fear and loathing among observers and theorists, but not so much anymore. Although astronomers may view the almost instantaneous emergence of structure in the early universe as a cosmological challenge, they don't necessarily consider it a dire threat to conventional wisdom. Far from being a bad case of regressive "groupthink," their confidence arises from increasing cross-pollination of ideas as well as close agreement by independent observers.

"We have the observational studies of metal-poor halo stars and we have the observational studies of the cosmic microwave background," says Tumlinson. "Until recently everyone would have told you those two had nothing to do with each other. Then you can also bring in the computer simulations of the high-redshift universe and the computer models of stellar evolution that we use to make predictions about how stars will evolve and look. So it's only by getting all these different people together that we've managed to do anything at all, and *that* has only been occurring for the last few years."

Grinning, Tumlinson adds a postscript: "It's not that we don't like each other; we just go to different meetings."

Add to these collaborative efforts the ongoing investigations of particle physicists—in particular, how stars create their thermonuclear energy—and you lay the foundations for new insights. As good as stellar models are, there is still much to be learned. In early 2004 researchers at the Laboratory for Underground Nuclear Astrophysics, located some 4,600 feet below the craggy limestone peak of Gran Sasso in northeastern Italy, successfully reproduced the carbon cycle of stars by colliding high-energy protons with nitrogen nuclei. They then measured how long it took for this reaction to produce oxygen and the simultaneous emission of energy. To their surprise, they found that the efficient carbon-nitrogen-oxygen cycle is not so efficient; in fact, it occurs two times more slowly than expected. When the new lower value for the rate of the carbon cycle is factored in, the age of the oldest stars in the Milky Way's globular clusters jumps from 13 billion years to 14 billion. If the Gran Sasso simulation holds up, then the universe, too, must be older by this amount. Perhaps stars had more time to form and build up into galaxies than we think.

The current polyphony between theory and observation may translate into beautiful music, but the magnum opus is far from over. Nevertheless, researchers are hopeful that given more observations with increasingly sensitive telescopes, and given better theories and more sophisticated computer models, the Dark Ages of galaxy formation will lift, perhaps within a decade, revealing the truth about the first cosmic egg.

6

IT STINKS, BUT IT ROCKS

Unless God created the universe with the limited computational ability of physicists in mind, its simplicity would seem to be very unlikely.

—B. J. Carr, *The Irish Astronomical Journal*, 1982

It is a fine April morning in College Station, Texas. The air is dry and cool, the sky remarkably blue. Indian blankets, scarlet paintbrushes, and bluebonnets in riots of color blanket highway medians and fields. At the Texas A&M University golf course at the corner of Texas Avenue and George Bush Drive, the greens seem especially vivid and verdant. One would be hard-pressed to imagine how any of the lucky players ambling down the fairways in their splashy golf togs could be quarrelsome in such temperate weather. This is a day for taking mulligans.

Meanwhile, in a windowless lecture room on the seventh floor of Rudder Tower, more than three dozen coffeed-up cosmologists are raptly listening to Max Tegmark describe his worst nightmares. Tegmark, a cosmologist at the Massachusetts Institute of Technology, looks like a younger version of Rolling Stones guitarist Keith Richards, complete with tousled hair, wry grin, and off-the-wall sense of humor. Known for his dry wit and complex but entertaining PowerPoint presentations, Tegmark is one of the standard-bearers of a new province of research called precision cosmology, in which theory and a slew of new observations and measurements are combined to constrain cosmological models and their parameters as never before.

But a few little problems in this precision paradise are giving Tegmark nightmares. "My nightmares," he tells his audience, "are that nature will refuse to give us any further quantitative clues about dark matter, dark energy, and the origin of the universe." The cosmologists are understandably sympathetic, and many no doubt have experienced similar night terrors. Tegmark stands before them now expounding upon the unthinkable: "If just one of my nightmares comes true, we'll have to look for other things to do."

Nightmare 1: Astrophysicists will find no more clues about dark matter. "No direct detection, no dark matter production [in linear colliders], and no astrophysical evidence for departures from 'vanilla' [perfectly cold] dark matter," Tegmark says. This would, indeed, be a predicament. Dark matter is invoked to explain how normal matter could so quickly have assembled into galaxies, galaxy clusters, and black holes in the very early universe. Of course, if dark matter is never detected, that wouldn't necessarily preclude its existence, but scientists' belief in it would be a matter of faith more than physics.

Nightmare 2: No further clues about inflation will be found and, just as bad, none about primordial gravitational waves. Inflation is thought to have created gravitational fluctuations, like the density perturbations apparent in the cosmic microwave background, in the fabric of spacetime. These fluctuations would have wavelengths as big as the visible universe, but so far none have been detected. No gravitational waves, no inflation, and no explanation for the size of the universe and its large-scale structure and uniformity—among other things. Cosmologists hope that future probes will detect these gravitational waves; otherwise all the air will be let out of inflation's balloon.

Nightmare 3: No more clues about dark energy will turn up. "This would be the end of the road for observational dark energy research," frets Tegmark. "It would leave us with no further quantitative clues about the nature of dark energy." Precision measurements of its density over time may yield nothing more than a flat-line constant or, in Tegmark's words, the "EKG of a dead cosmologist."

This is indeed a distressing prospect. From the Type Ia supernovae findings as well as data from WMAP, the Sloan Digital Sky Survey, and the 2-degree Field Galaxy Redshift Survey, astronomers have been able to pin down the density of dark energy—nearly three-quarters of the universe is made up of the stuff—but nobody knows whether the density of dark energy changes as the universe expands. That's a problem, say, if you want to tie in the formation of galaxies and galaxy clusters with the dynamic history of the universe. Is it indeed constant? If so, then dark energy may simply be Einstein's built-in cosmological constant, which corresponds to a constant energy density and a constant pressure exerted by goings-on in the quantum vacuum (forgetting for the moment that physicists' theoretical predictions yield a cosmological constant that is 120 orders of magnitude higher than the observational value). If it is not constant, then bring on quintessence, with a negative pressure that varies over time, or some new form of wild physics that modifies Einstein's theory of gravity.

The situation has forced Tegmark to leave no stone unturned

in the quest for dark energy. "Since we don't know what dark energy is," he tells his attentive colleagues, "I went to my source of all knowledge—Google—and typed in 'dark energy' and clicked under the image search. This is what I learned." He presses a button on his computer's remote control, and a picture of a white quart-size bottle emblazoned with the label *Dark Energy* flashes onto the screen. Below the label is a tagline that Tegmark savors as he reads it aloud: "It stinks, but it rocks."

Apparently "Dark Energy" is a plant nutrient developed by a company called American Hydroponics. The owners of the company might never have imagined their product being used as a visual aid in a cosmology lecture. Still, the label is appropriate enough. Dark energy stinks in that nobody knows what it is, but it rocks in that it helps explain an otherwise baffling finding—that cosmic expansion is not slowing down but accelerating. But just as important is the now-accepted view that without dark energy and its compadre in gloom, dark matter, the universe we know wouldn't exist. All the large-scale structure we see in the universe might never have had the chance to form and the universe wouldn't have attained its present size. In fact, instead of expanding, the universe might just as well have collapsed back upon itself long before Earth became a living planet. There must therefore be some force acting on the cosmos at large scales as a supplemental impetus to ordinary expansion. If not—well, there ought to be. To borrow from Voltaire, if dark energy did not exist, it would be necessary to invent it.

Beginning in 1998, studies of a particular class of very distant exploding stars, called Type Ia supernovae, revealed an anomaly when nearby Type Ia's were compared with those lying at much greater distances: the distant Type Ia's weren't as bright as they should be. For over a decade prior to this finding, astronomers had labored to reconcile the range of luminosities and energies emitted by Type Ia supernovae in order to use them as "standard candles" for measuring distances across the universe. After much theoretical toil and observational sweat, they thought they had succeeded, but now it seemed they were faced with an unforeseen muddlement.

In the 1940s and throughout the 1950s, astronomers recognized only two broad classes of supernovae—Type I and Type II—the identification of which, by Fritz Zwicky in the 1930s and by German-American astronomer Rudolph Minkowski in 1941, was considered a great triumph for stellar evolution. Type II supernovae displayed a diverse range of peak brightnesses, but they were typically found in the star-rich arms of spiral galaxies and their spectra indicated the usual abundances of metals and hydrogen. For these reasons, astronomers assumed that the progenitors were massive stars that had undergone a complete core collapse. Type I supernovae, on the other hand, had a smaller range of peak brightnesses. Their ejected gases expanded more slowly than Type IIs, and their spectra showed no hydrogen at all. This finding, along with the fact that they were the only type of supernova seen in ancient elliptical galaxies, suggested that Type I supernovae had undergone extensive evolution, billions of years longer than Type IIs, and were thus sapped of their hydrogen fuel. The likely Type I progenitor was therefore a white dwarf that had gravitationally slurped up gas from an orbiting companion star, until the accumulating mass triggered a thermonuclear explosion.

But by the early 1960s and extending into the early 1980s, it dawned on astronomers that not all Type I supernovae were identical. In a 1964 paper Italian astronomer Francesco Bertola was the first to give the oddballs some class distinction, calling them "Peculiar Type I" supernovae. As observations progressed, it became clear that Peculiar Type I supernovae consisted, in fact, of at least two separate classes of exploding stars. One class exhibited silicon in its spectra and produced a lot of radioactive nickel during its nuclear conflagration. A second class lacked silicon, produced fewer radioactive by-products, and was not as luminous. Astronomers christened the first class Type Ia. The second was divided into two subclasses, Type Ib and Type Ic. (Not to belabor the astrophysical detail here, but it turns out that the progenitors of Type Ib and Ic supernovae, though they are hydrogen-free, are thought to be related to Type IIs in that they were massive stars that had undergone core collapse and had exploded in the arms of spiral galaxies. For better or worse, astronomers base supernovae types on how their spectra appear at maximum brightness, never mind the trigger mechanism.)

By recognizing these unique spectral signatures, astronomers were able to look for consistent traits among Type Ia supernovae and to cull those whose peak brightnesses, colors, and light curves differed significantly from the norm. This work elevated Type Ia supernovae to one of the most reliable standard candles in cosmology. Moreover, with a luminosity equal to 4 billion suns, Type Ia supernovae were bright enough to be seen across billions of light-years of space. Given the improvements in assorting them into a group of like kind, by the mid-1990s astronomers were confident that they could calibrate their Type Ia models with real-world observations of redshifts and magnitudes. The results would reveal for the first time how much expansion had slowed since the universe was young. Things were looking up. No wonder, then, that cosmologists were caught flat-footed when their observations pointed to something other than what their hard-won models had assuredly predicted.

For years astronomers have assumed that expansion has been

slowing due to the gravitational drag of all the matter and dark matter in the universe. Hence the prediction was that when telescopes finally captured a Type Ia supernova from a considerable cosmological distance, it would exhibit a gradual and predictable downward trend in brightness with increasing redshift. In other words, a decelerating universe would convey observers at an increasingly slower rate, which in turn meant that a supernova's peak brightness would decline more gradually with distance. But as astronomers began racking up brightness measurements of dozens of high-redshift Type Ia supernovae, they found something else. At a redshift distance of 0.5, supernovae were some 10 to 20 percent *fainter* than astronomers expected in a leisurely decelerating universe. That meant we were being carried away from these remote supernovae at an increasing rate. The universe wasn't slowing down as everyone had thought; it was accelerating.

How was this possible? What force could counter gravity? A lot of weird ideas were proposed—and still are—but only one regularly outweighs the others: Albert Einstein's cosmological constant in the incarnation of a new form of energy—*dark* energy. Perhaps it hadn't been such a blunder after all.

Today acceleration has been confirmed from measurements of hundreds of distant Type Ia supernovae by different research groups, including the Lawrence Berkeley National Laboratory's Supernova Cosmology Project, led by Saul Perlmutter, and the Hubble Higher-z Supernova Research Team, led by Adam Riess. In fact, it was Riess's group that first dispatched the astonishing news to the astronomical community.

"That was a total surprise," Riess recalls. "And nobody knows what dark energy is. It may have to do with totally different physics, or it may be related to our lack of understanding of how gravity really works on large scales in the universe."

Whatever *it* is, cosmologists are now forced to deal with the idea that 70 percent of the universe consists of stuff that stinks but rocks. All the while observers continue to seek out Type Ia supernovae to probe ever deeper into the cosmic sediments, reaching redshifts closer and closer to 1, or nearly 7.5 billion years in the past.

Obviously, a lot is riding on the conclusions astronomers have drawn from Type Ia supernovae. A fair question to ask, then, is how astronomers know that the ones seen in younger galaxies are the same beasts we see in nearby galaxies. After all, the universe was a different place 7 billion years ago—smaller, denser, and more metal-free. Might these factors affect the intrinsic properties of more distant Type Ia supernovae and hence falsify the findings?

As of this writing, that prospect seems doubtful. For a while it was argued that the luminosity of higher-redshift supernovae could be more highly attenuated by billions of light-years of intergalactic dust than by those that are not that far away. But normal dust doesn't just dim—it also reddens light in ways that can be quantified. Assuming that the dust in distant galaxies is similar to the dust in our own, astronomers can correct for this kind of reddening. So far they have found that distant Type Ia supernovae are not significantly more obscured than local ones.

An even bigger question is whether these exploding stars might be compositionally different from proximal Ia supernovae and thus yield different luminosities. But even this challenge has lost ground. At the same Texas A&M symposium where Max Tegmark discussed his cosmological nightmares, Adam Riess delivered a teleconference lecture from the Space Telescope Science Institute. In the audience sat Craig Wheeler, a savvy veteran supernova expert with the University of Texas at Austin. In the mid-

1980s Wheeler was among the pioneering researchers who identified Type Ib supernovae as a new subclass. At one point during his presentation, Riess remarked that his group saw "no chemical differences between Type Ia's," to which Wheeler ever so slightly shook his head. At a break in the talks, I asked him if his apparent dissent meant that he did not agree with Riess's assessment. He shrugged and smiled, preferring to remain politic for the moment.

In a later e-mail response, Wheeler was more forthcoming. It's not that he disagrees with Riess, but rather that he is mystified. Why don't distant Ia supernovae appear different from those that are closer? "On average, the Ia's at redshift of 0.5, where the accelerating universe was discovered, look very much like those 'nearby,' " says Wheeler. "We are still trying to understand why that is true. There is a little scatter in the brightness, about fifteen percent. That must be where the expected differences lie, but nothing systematic has been teased out [of the data]. Every supernova person in the world is hammering on every new event to try to compile the statistics that would reveal such effects."

Wheeler points out that if observations and various independent tests continue to support the idea that distant Ia supernovae behave the same as those nearby, then it's high time to develop better physical models. "A point to which I am rather sensitive," he says, "is that the [recent Type Ia] supernovae were used as purely empirical tools. No 'understanding' was required at all. In the next phase there is a general feeling that we must understand them in the sense of physics to get a handle on the systematic effects . . . before we can do the really precise work to measure the equation of state of the dark energy. But I have also heard people . . . argue that even the next stage will best be done purely empirically, that is, independent of any physical 'model.' We'll see."

If we accept that dark energy is real, what is the next logical step? Riess wants to use the supernova data to pin down the properties of dark energy—particularly its pressure, or strength, versus its density, or pervasiveness, throughout the universe. This relationship, cited by Wheeler above, is called the equation of state, and as we shall see, finding it is one of the leading crusades among astrophysicists today. The best way to establish the equation of state of dark energy, says Riess, is to continue observing distant Type Ia supernovae to probe different periods in the history of the universe. "We've been increasing our knowledge of that rapidly in the last few years. Just collect more of this data, and our precision in measuring dark energy will improve." Then his face darkens. "But we won't be able to do this kind of research without the Hubble Space Telescope. We'll have to do something else."

For now, supernova researchers still rely on correlating magnitudes with redshifts, but they don't need to obtain the largest redshifts. Acceleration becomes apparent out to redshifts of only about 0.5, which corresponds to an epoch about 5 billion years ago. At redshifts above 1, or 7 billion years ago, the supernovae actually brighten again as they would in a nonaccelerating universe. "We don't want to go to very, very high redshifts with supernovae," says Riess, "because dark energy only became important in the last five or seven billion years in the history of the universe. [That's when] dark matter let go of the universe, allowing it to accelerate."

That the transition from expansion to acceleration happens to coincide with the formation of our solar system hasn't escaped the notice of physicists. Many consider the overlap too much of a coincidence—anthropical, to be more precise. The concept abounds with coincident parameters that favor human existence—not just the expansion rate of the universe but physical constants and interactions such as the weak and strong nuclear forces, electromagnetism, and the gravitational constant. Moreover, a just-right universe requires humankind to play a more exalted role in existence than that of passive observer. Why, detractors ask, should the cosmos seemingly depend so much on our being here to observe it?

Instead, they prefer to seek alternative interpretations, such as the one proposed by Gia Dvali, Gregory Gabadadze, and Massimo Porrati of New York University. In their theory, called the DGP model, universal acceleration can be explained by amending Einstein's general theory of relativity and adding a fifth dimension. For those who may need a reminder, general relativity was originally proposed to explain how matter affects space and time. Einstein concluded that the gravitational fields surrounding matter change the geometry of spacetime by curving it. The greater the mass, the greater the curve.

Hold on to your theoretical hats. In the DGP scenario, gravity is Newtonian only at short distances, up to 100 million times the size of the solar system. At distances beyond that, a crossover occurs where gravity can stretch out and even escape our four-dimensional universe. It does so by leaking gravitons—theorized particles that carry the gravitational force—from our universe, which the theorists refer to as a membrane, or just "brane," into a fifth-dimensional brane. As the gravitons evaporate, gravity at large scales gradually loosens its grip on matter, and the universe becomes naturally self-inflationary.

Still another idea, proposed by physicist Ann Nelson and her colleagues at the University of Washington, is that dark energy could be the result of interactions between neutrinos and a previously unrecognized subatomic particle they call the "acceleron." Their theory requires no extra dimensions, but it is nonetheless a theoretical spin on how events at micro scales can have macro repercussions.

Nelson contends that dark energy is the result of the universe's trying to pull neutrinos apart. Neutrinos are created by the gazillions in the nuclear furnaces of stars, including our Sun. Like tiny ghosts they flit throughout the universe, passing through all matter, including people. But because neutrinos have negligible mass and no electric charge, they interact very little, if at all, with the materials they penetrate. The hypothetical accelerons have even weaker interactions with matter, but they can couple with neutrinos, causing them to undergo a change in mass. As the universe tries to pull them apart again, a tension is produced like that of a stretched rubber band. That tension, says Nelson, fuels the expansion of the universe.

One of the intriguing aspects of the acceleron theory is that it predicts that acceleration will gradually slow until it ceases altogether. "In our theory," says Nelson, "eventually the neutrinos would get too far apart and become too massive to be influenced by the effect of dark energy anymore, so the acceleration of the expansion would have to stop. The universe could continue to expand, but at an ever-decreasing rate." That turnover, she says, should happen sooner than later. "Acceleration should stop relatively soon, on cosmological scales, now that the interneutrino spacing is already becoming somewhat large compared with the dark energy scale—roughly a millimeter, which should be the range of the force."

Of course, the problem with these alternative theories—never mind that none really dispatches the anthropic coincidence—is their lack of verification. In the words of one noted astronomer, the recent plethora of unconventional notions cited to explain dark energy and dark matter are just "ornamental, intellectual games."

In any case, it appears that acceleration and dark energy are digging in for a lengthy, if not permanent, stay. During a symposium at the Space Telescope Science Institute in May 2004, John

Tonry, a cosmologist at the University of Hawaii, put the study of dark energy into succinct perspective: "We're not so interested in whether the universe is accelerating or not—we *know* it is accelerating. The question is, why?"

Which returns us to the question of whether the dark energy density is constant or varies throughout the universe. This is important. If it varies, it could increase in the future, thus rending the universe asunder right down to its last atomic particle, or it could decrease, essentially handing the ball back to gravity and allowing the universe to collapse or at least decelerate. Current thinking, particularly in light of the WMAP findings, favors an unchanging density in a flat universe, reviving the terms of Einstein's original cosmological constant. The source of acceleration in this case wouldn't be branes or accelerons but virtual particles. One might call this the "seltzer effect."

Quantum theory holds that the vacuum of space teems with particles that continuously pop into and out of existence like soda bubbles. In so doing, they collide with one another, annihilating themselves and their assorted antiparticles, creating a subatomic effervescence. Given the size of the universe, these quantum-sized bursts of energy, each thought to last no more than 10^{-21} second, would endow the vacuum with a net density, which in turn could provide the necessary propulsion to drive acceleration as the universe expands and matter thins to lower densities.

Unfortunately, in addition to the cosmic coincidence mentioned earlier, a vacuum energy creates another conundrum for cosmologists. Physicists working on quantum field theory have long held that the vacuum energy might contribute as much as 70 percent to the matter-energy budget of the universe. It is, in fact, one of the natural outcomes of their theoretical calculations dealing with the lowest energy state of particles, the gist of which is that even an apparently empty vacuum is not totally devoid of energy. But their calculations predict that the vacuum energy should be between 10^{50} and 10^{120} times *greater* than the value derived by cosmologists' observations. If it were as great as the physicists' calculations say it is, the universe would have expanded itself right out of existence long, long ago.

Perhaps, then, there is some sort of unknown mechanism at large in the universe that almost exactly cancels the vacuum energy. If so, what would it be? Paul Steinhardt, a Princeton theorist whose research spans particle physics, astrophysics, and cosmology, proposes a different form of vacuum energy, which he calls "quintessence." Unlike the cosmological constant, which never changes in density even as the matter density of the universe diminishes, quintessence declines as the universe expands. Moreover, because the vacuum energy is thought to evolve over time, perhaps it's not a coincidence that the dark energy density of the universe, which is 0.7, is close to that observed for dark matter, which is 0.3. Some theorists would respond, however, that building an evolutionary component into quintessence does not necessarily dodge the cosmic coincidence problem. It is still guilty of fine-tuning and, some might again say, of being just another ornamental intellectual game.

Maybe cosmologists are barking up the wrong theoretical tree altogether. Perhaps dark energy and dark matter are hermaphroditic in nature, the yin and yang of the cosmos. This somewhat mystical notion has been given voice in a number of quintessence-like theories that combine dark energy and dark matter into a single component. One, called unified dark matter (UDM) or "quartessence," arises from colliding brane worlds. Quartessence explains the acceleration of the universe by invoking an exotic equation of state that makes it behave like dark matter at high density and like dark energy at low density.

Perhaps because cosmic acceleration seems so exotic, theorists tend to offer exotic explanations by invoking new or unknown ingredients (exotic particles) or physical conditions (cosmic strings and leaky gravity). But what if acceleration is the result of something that has already happened, some natural property of the universe that doesn't require new physics? Something that, whatever it was, did its thing and split but left effects that we can still see and measure today? Like Elvis, it has left the building, but it still looms large in the universe. This no-dark-energy-required approach forms the basis for an alternative explanation suggested by a group of theoretical physicists from the United States, Canada, and Italy and has two of the qualities scientists most admire: it's simple and it's elegant.

Edward W. Kolb of the Fermi National Accelerator Lab in Chicago, Sabino Matarrese of the University of Padua, Alessio Notari of the University of Montreal, and Antonio Riotto of the National Institute of Nuclear Physics in Padua propose that cosmic ripples in spacetime, generated by the sudden outward gust of inflation, can account for both expansion and the acceleration of the universe. Today these ripples have been stretched by expansion, and they now extend well past our cosmic horizon, a distance of over 15 billion light-years. As these superripples continue rushing outward, however, they heave the universe along with them. It's as if the cosmos were being carried along on a wave—or more accurately, says Kolb, it's as if "we are riding a trough, not a crest of the wave." Moreover, he says, acceleration won't moderate in the future: "The sign of the effect will not change with time. We will always accelerate."

If expansion *is* more rapid today than it was in the past (which is implied by the very definition of the word *accelerate*), then a more sluggish inception could help explain how galaxies assembled so rapidly in the early universe. Recent observations by astronomers from all quarters have turned up massive galaxy clusters in place when the universe was between 1 billion and 2 billion years old. These could have assembled well before gravity passed the baton to dark energy, which, as stated by Adam Riess, was about 5 billion years ago. Before that time, the universe had a much smaller volume—over three hundred times smaller at a redshift of 6 and a thousand times smaller at a redshift of 10. Matter was thus closer together, and gravity could have swept galaxies out of the gaseous background like cotton candy onto a paper cone. Later, as the universe expanded and matter further separated, gravity would begin yielding to the repulsive effects of dark energy, making it more difficult for galaxy clusters to amass. Thus it makes sense to search for the effects of dark energy in galaxy clusters at various redshift distances to determine the rate of clustering over time.

Astronomers with the Sloan Digital Sky Survey (SDSS) did just that by correlating millions of galaxies reaped in their survey with the cosmic microwave background (CMB) temperature maps from WMAP. In this approach the CMB provides the photons and the galaxy clusters provide the gravitational wells for those photons. In a universe consisting of normal matter and no dark energy, microwaves falling into a cluster's gravity well would first gain energy, like skiers hurtling down a steep slope, and thus appear hotter. When they exited, they would lose that energy, like skiers trying to reach the top of the next slope using their acquired momentum, and consequently appear cooler.

But that's not what the SDSS astronomers saw. They found

that the microwave background is slightly hotter in regions containing galaxy clusters. As the CMB photons pass through the gravity wells of these clusters, they appear to receive an extra "kick" out of the well. The kick is caused by dark energy supplementing expansion and thus making the gravitational wells shallower. Hence photons falling into overdense regions like galaxy clusters gain more energy than they lose when escaping the gravity well. The net amount of the gain is small, less than one part in a million, but it is enough to show that the effects of dark energy can be detected at mass concentrations only 100 million light-years across, one-hundredth that of previously observed effects.

These observations indicate that dark energy's influence can be measurable down to at least galaxy cluster scales. But even if dark energy is also present at the scale of individual galaxies, as is suspected, its sway may be less important than that of dark matter. Astronomers pin their understanding of how the first galaxies assembled themselves on the existence of dark matter, specifically *cold* dark matter. The stuff making up cold dark matter is thought to be exotic elementary particles, such as axions, WIMPs, or closed cosmic strings (as opposed to the open variety). Whatever it is, as with dark energy only the indirect gravitational effects of dark matter can be observed. And like dark energy, it, too, stinks and rocks. It stinks in that no one knows what it is truly made of, but it rocks because it resolves all manner of problems. It explains why the rotations of spiral galaxies remain constant in their outermost disks and why galaxy clusters don't fly apart, as a tally of their visible mass suggests. And as mentioned, it's an absolute requirement if you need to explain the rapid buildup of galaxies, galaxy clusters, and galaxy filaments and sheets in the early universe.

What if, as Tegmark worries, dark matter is never found? This prospect gives cosmologists the shudders, because if dark matter fails to turn up in an accelerator or a neutrino detector one of these days, the null result will be the cosmological equivalent of a Wall Street crash, or worse.

As in the case of dark energy, "all we know about dark matter," Tegmark says, "is its density." It would indeed be nightmarish for cosmologists to have groped their way through the shadows only to realize that their dark matter and dark energy paradigms have led them into an observational quagmire from which there is no escape.

Despite his nightmares, Tegmark doesn't really think they will come to pass, mainly because of the strong concordance arising from diverse lines of research. "The reason I feel so confident in this field today is because different teams are coming up with the same answers. It didn't have to come out this way," he says, pausing for drama's sake, "but it did."

Tegmark refers to the dozen or so observational criteria called cosmological parameters that describe the expansion rate, acceleration, geometry, and energy and mass budgets of the universe in mathematical terms—all of which contribute to the new field of precision cosmology. Observers may use different approaches to derive exacting numbers for these cosmological parameters, but the important thing, Tegmark says, is that the values closely agree. The Hubble Constant, the rate at which the universe expands, is just one example. "We used to argue about the Hubble Constant," he says, referring to the contentious days when estimations of it varied by a factor of two. "Now we get complete agreement using independent analysis."

Princeton cosmologist P.J.E. Peebles concurs, though he prefers the term *accurate* to *precision* cosmology. *Precision* refers to the number of significant figures quoted in the value of a measurement, he says; *accuracy* refers to the number of figures that remain significant after one takes account of systematic errors.

"To be confident about our deductions of aspects of fundamental physical reality from limited observations of processes operating on the far side of the visible universe," Peebles says, "we need to demonstrate consistency of evidence from a broad variety of astronomical methods that are subject to quite different uncertainties. The remarkable thing is that this can be done, at least for some simple purposes. The consistency of estimates of the mean mass density based on a broad variety of methods shows we almost certainly have not been misled in concluding that the mass density is well below the Einstein–de Sitter value."

For all this accord, the fact remains that as limitless as the universe appears, it is vastly more limitless than we know. The situation, says John Tonry, "is equivalent to not knowing what water is even though it covers three-quarters of the Earth's surface." In a February 2004 address at Cambridge University's Institute of Astronomy, noted cosmologist Sir Martin Rees even conceded that "it is embarrassing that 95 percent of the universe is unaccounted for."

Fortunately, ignorance is not tantamount to folly. Not knowing the physical properties of dark matter or the staying power of dark energy, at least for now, may be less important than understanding how they may have twiddled the knobs of the cosmic machine. New and more powerful telescopes on the ground and in space promise to provide more precise measurements at large scales, while particle accelerators, neutrino telescopes, and cryogenic detectors may produce the first definitive evidence of dark matter particles—perhaps a new class of heavier, slower-moving "partners" that shadow every particle known to exist. The search for dark energy will be an even greater challenge, but that hasn't prevented physicists from proposing ways of looking for it.

If the research momentum of recent years is any indication, breakthroughs may occur sooner rather than later. It would be unwise, however, to be overly predictive of the outcome. Optimism may rule the day, but not without a prudent measure of reserve. The universe truly is not as it seems. In fact, it might even be said that given the unknowns of dark matter and dark energy, observational cosmology has failed to a level of 95 percent to describe what the universe is made of. Perhaps the nightmares of cosmologists are not entirely groundless and some of the vocal consensus nothing more than bravado. As Freud once wrote, "When the wayfarer whistles in the dark, he may be disavowing his timidity, but he does not see any the more clearly for doing so."

7

KECK: SAILING ACROSS AN UNKNOWN OCEAN

Shall we consider the universe as extending into infinity and conclude that bigger and better telescopes will always reveal to the inquiring eye of an astronomer new and hitherto unexplored regions of space, or must we believe, on the contrary, that the universe occupies some very big but nevertheless finite volume, and is, at least in principle, explorable down to the last star?

—George Gamow, *One, Two, Three . . . Infinity*, 1947

Fifty thousand dollars a day, or about a dollar a second. That's how much it costs to run the largest telescopes in the world, the 10-meter Kecks I and II atop Mauna Kea, Hawaii. Of course, that's nothing compared to the Hubble Space Telescope, which has been estimated to cost twice that per day. But of course one doesn't have to worry about cloud cover, snow, or excess humidity in space. For Keck, rain or shine, win or lose, it's still fifty grand.

No one knows this better than Richard Ellis, who has spent many a long night observing at Keck trying not to waste his time or the observatory's money with instrument glitches or foul weather. Admittedly, some of those nights were unavoidable busts due to one or both reasons, but in Ellis's case, the wins have more than made up for the losses. His observations of the distant universe have contributed much to what astronomers know about how galaxies form and assemble into galaxy clusters. In the last five years alone, he and his research team have racked up twelve galaxies with redshifts of 5 and above, two of which were the subject of journal papers. They fully expect to find more, perhaps ones with redshifts as high as 9 or 10. As mentioned in Chapter 2, Ellis's most impressive results in this endeavor to date have been obtained using gravitational lenses around massive foreground galaxy clusters to ferret out galaxies in the background. These warped regions of space act as a supplementary magnifying lens for the telescope, allowing observers to peer through it and see objects that lie even further back in time.

In the summer of 2003 Ellis's team discovered a redshift 6.8 galaxy. One year later they returned to Mauna Kea to try their luck again. Ellis was generous enough to invite my wife, Alexandra, and me along, and we happily took him up on his offer. Alex, a science journalist herself, wanted to write about how astronomers go about searching for the most distant galaxies. There's no place better to see how it's done than Keck.

When we arrive at Waimea, a bustling, crowded town in the foothills of Mauna Kea, it is raining and the nearby mountains are draped in scudding tattered gray clouds. A flash-flood watch has been issued for all of the Big Island. The weather certainly looks grim from where we are, but the Keck telescopes are removed another two miles higher in elevation, where the arid air rules and the clouds are capped by an atmospheric inversion layer. Up there, above the murk, the Sun is shining. We hope.

The following day, the day of the observing run, the weather somewhat improved, we head out for the summit of Mauna Kea.

Our purpose in going to the top is to have a look for ourselves at the massive telescopes and instruments used to search for the youngest, most distant galaxies in the universe.

Of course we have to get there first, and the only way to do that is via the notorious Saddle Road, which winds for fifty-three miles between Mauna Kea and Hawaii's other grand volcano, Mauna Loa. This is an adventure unto itself, but having visited Mauna Kea a few years earlier, we know what to expect. A little background here. The road was hastily carved out of the lava-encrusted surface in 1942 by the U.S. military in order to link up the east and west sides of the Big Island. They were successful, but the road, as is said of anything of dubious quality, is a piece of work. It is the only state highway in the United States that is considered off-limits to rental cars. This is something Alex reminds me of as we climb into our own rental death trap.

Even though significant improvements to the road have been made in recent years, it's still a white-knuckle drive all the way to the Mauna Kea turnoff. The two-lane blacktop snakes through mile after mile of slaglike lava flows, cinder cones, and scrub country, punctuated here and there by astonishing islands of blossoming foliage. Unbanked blind curves, one-lane bridges, and sudden dense fog make it all the more thrilling. Then there's the Pohakuloa Training Area, through which Saddle Road threads. Along this stretch you're liable to see camouflaged tanks, Humvees, and other military vehicles paralleling you or crossing ahead. Alex takes delight in pointing out the ominous signs that are posted here and there warning, "Danger! Live ammunition impact and dud area," and "Danger! Live ammunition firing overhead." Supposedly, if you pull over and listen carefully, you can sometimes hear the distant booms of exploding artillery. We decide not to do this.

Despite the harrowing drive, we make it to the observatory

turnoff in one piece. After a quick lunch at the chalet-like as-tronomers' quarters known as Hale Pohaku, nine thousand feet up the southern flank of Mauna Kea, we wend our way to the top over a teeth-rattling washboard switchback named John Burns Way af-ter the governor of Hawaii who, in 1984, authorized the road's construction. Our tour guide for the afternoon is Peter Michaud, a public information officer for Gemini Observatory, which is also atop the mountain and is one of the foremost telescopes being used to plumb the undiscovered depths of the early universe. Leaving most of the clouds behind us, we head up into ethereal ul-traviolet skies. We stop only once along the way so Alex, whose de-gree is in geology, can explore an exciting mound of gravelly glacial till. As she happily examines the grayish rubble, I check the air temperature. It's hovering around 50 degrees Fahrenheit, quite balmy for this altitude. When we climb out of the car at the base of the Gemini dome five minutes later, we find that it has dropped a mere 10 degrees, still very pleasant sweater weather. From up here, the sky is the deepest blue and the view is stunning beyond words. But it is the quiet, the absolute stillness, that I find the most profound. Any sound, a puff of wind, a voice, the crunching of gravel, or the swishing of jackets, seems amplified by the silence. Until we are in a quiet place, like the top of a mountain, we don't realize how much babble and noise there is in the world.

Michaud, a fount of information about Mauna Kea's history, in-forms us that in fact no telescopes occupy the true summit of the extinct volcano. "It always irritates me a little when I read that the telescopes on Mauna Kea are located *at* the summit," he says. "That's not true. There are no telescopes *at* the summit." He points toward a footpath leading up the flank of a lone cinder cone almost due south of where we are standing. Some sort of white structure can just be seen standing out against the blue sky at the top. "That's the true summit. There's an altar up there," Michaud tells us. For many native Hawaiians, the summit of White Mountain, as they refer to it, is one of the most sacred places in all the islands. Some, he adds, even scatter the ashes of their dead there.

The Gemini Observatory may not stand at the true summit, but that hasn't hampered its effectiveness by any means. It is one of two 8-meter telescopes actively digging deep into the universe. Its southern counterpart is located on nine-thousand-foot Cerro Pachón in the Chilean Andes. The two telescopes are unusual in that their mirror surfaces are coated with a thin layer of silver rather than aluminum. "The silver provides good reflectivity for one year before a new coating is applied," says Michaud. The main advantage of a silver coating is that it reduces thermal emission from the telescope, which in turn increases the sensitivity of the mid-infrared instruments. The Gemini telescopes also employ a sophisticated adaptive optics system that corrects for atmospheric distortions of starlight on the fly. These innovations make both Gemini telescopes ideal for studying distant galaxies, not to mention probing galactic phenomena, such as star-forming regions and the warm dusty environments encircling established stars, which may harbor young solar systems or extrasolar planets.

An hour later Peter drops us off at the Keck double domes, where we conscript Rich Matsuda, Keck's electronic engineering manager, to be our guide. The Kecks are truly the big boys of the mountain. Each of these telescopes comprises thirty-six mirror segments to form a total light-gathering surface 10 meters in diameter. The Keck telescopes also employ a cutting-edge adaptive optics system that uses a laser as an artificial guide star. This allows adaptive optics to be used across the sky, not just in those fields that contain bright stars, as was once the case.

The Kecks are perfect galaxy-finding machines, but not solely because of their superior light-gathering ability. They also boast an impressive array of sensitive, and expensive, light-gathering detectors that can be coupled to the telescopes. Matsuda guides me up a steep flight of blue steps over to the Keck II focus and pulls back a black Velcroed cover revealing one of these instruments, called the Near-Infrared Spectrometer, or NIRSPEC. This is the instrument Ellis will be using tonight for his observing run. It is not pretty by any means, but from an engineering standpoint it is a beauty. It consists of a red anodized aluminum box about the size of a small refrigerator lying on its side and mounted atop a steel support frame ensnarled by a mass of wires, cables, and insulated hoses. The red box is, in fact, a kind of refrigerator or, more precisely, a dewar. To obtain the spectrum of a distant galaxy shining at near-infrared wavelengths, NIRSPEC's detectors must be continuously cooled to a temperature of about 60 kelvins. As I stand there, I can hear ticking whenever the coolant kicks in.

Midway down the support frame, Matsuda points out four fiber-optic cables, color-coded white, gray, black, and red, plugged into a narrow aluminum faceplate. When incoming photons from galaxies strike NIRSPEC's detectors, they are converted into electronic signals that are then sent through these cables to the observatory's computer room, where they are digitally recorded.

Matsuda next takes us into the Keck control room, where Alex snaps pictures of me sitting at the main control station, my expression glazed somewhat by a creeping anoxia. The station is really just a long worktable that holds a bank of four flat-screen computer monitors and the usual desk debris, including a roll of toilet paper. Directly to my right is a large telemonitor with a squat black camera mounted on top. On the screen is a view of the remote observing room in Waimea. From there Ellis and his team will maintain constant communication with the telescope control technician who, in a few hours, will be sitting where I'm sitting now.

Matsuda next takes us into the computer room, with its towers of processors and instrument connection panels, but by then the altitude is finally getting to us. We've been climbing around the innards of the world's most sophisticated telescopes for nearly two hours. Alex, who has misplaced her sunglasses, frantically searches her jacket pockets but to no avail. Obsessed with finding them, she wanders around the room looking on desktops and computer stations, thinking she may have absentmindedly dropped them there. She will later find them tucked into one of her jacket's small zippered pouches, which she thought she had searched earlier. Meanwhile my eyes are feeling dry, and my peripheral vision is encompassed with halos. I can't vouch for my mental acuity, but I assume that not all of my brain's spark plugs are firing. It is clearly time to leave. Thank goodness Peter is driving.

When we step outside, we notice immediately that the weather has deteriorated. What had been a sunny and brilliant afternoon has been replaced by a cold rain and shrouds of clouds so thick they obscure the views of the other domes. Michaud picks us up, and we are soon trundling back down John Burns Way into the welcome oxygen.

On the drive back to Waimea, during which we encounter still more rain and fog, plus two Saddle Road car crazies taking curves at ballistic velocities, it occurs to me that Ellis and his team are fortunate in not having to negotiate "Purgatory Road" every time they observe at Keck, much less adjust to the thin atmosphere atop Mauna Kea. In fact, most astronomers haven't had to observe at the top since 1996, when the remote observing facility in Waimea was established. Today most of them fly into Kona Airport from the mainland, drive to Waimea, and ensconce themselves in the comfy visiting scientist quarters, where they crash during the day. In the early afternoon they convene in one of two Keck observing rooms to begin preparations for the hard day's night that lies ahead.

Ellis is there now with his team. When I phone him, he reports that they are making sure everything is in readiness before we arrive. "We've not used this instrument [NIRSPEC] for this project before," he says, "so this is new territory." The plan, assuming we survive Saddle Road and everything goes smoothly in the Keck control room, is to meet Ellis at eight o'clock at Keck headquarters on Mamalahoa Highway. After that the only other concern is the weather, which, Ellis warns, looks uncertain.

At the appointed hour Alex and I arrive at Keck headquarters, noted for its impressive glass facade patterned after the telescopes' segmented mirrors. Ellis welcomes us with a cheery "Hello, hello" in his lilting Welsh accent. He wears white khaki slacks and a light blue shirt emblazoned with canoes, fishes, and other aquatic forms. His smile is amiable and the conversation casual, but it is apparent by his brisk pace back toward the complex that his thoughts are fully on the prospects for tonight's observing run. After passing down a long, narrow corridor, we arrive at the Keck II control room. Altogether, it doesn't seem much larger than four office-style cubicles, but nonetheless it's arrayed with a sagging, well-worn sofa and a horseshoe arrangement of worktables topped by a dozen computers and notebooks. Ellis's two research team members, Michael Santos and Daniel Stark, sit expectantly in brown swivel chairs, a scatter of papers, photographs, and empty pizza boxes before them. Santos, who has the black hair and beard of a Canadian lumberjack, is a former grad student now working as a theorist at the Institute of Astronomy in Cambridge, England.

THE ULTRA DEEP FIELD

The Hubble Ultra Deep Field records at least ten thousand galaxies brimming over a region of sky one-tenth the apparent size of the full moon. The smallest, reddest galaxies may be among the most distant, existing when the cosmos was less than 800 million years old. The ruddy-hued spheroid galaxies contain mostly old stars, while the blue-colored disk galaxies are still in the process of forming stars. Almost all of the galaxies in this image have unusual shapes, indicating that they exist at an earlier stage of cosmic evolution than those we see in the nearby universe. (NASA, ESA, Steven Beckwith [STScI], and the HUDF team)

H U D F   D E T A I L S

Close-ups of select fields in the HUDF reveal the bewildering shapes of young galaxies. Many, such as the blue wisps in the middle left panel opposite, have been fragmented by multiple galaxy collisions. A large disk galaxy seen edge on appears in the bottom left panel as a bright streak. The beautiful face-on spiral in the bottom right panel is an isolated galaxy. At a distance of one billion light-years, it is also one of the nearest in the HUDF image. The bright orange object in the middle right panel is a nearby star. (NASA, ESA, Steven Beckwith [STScI], and the HUDF team)

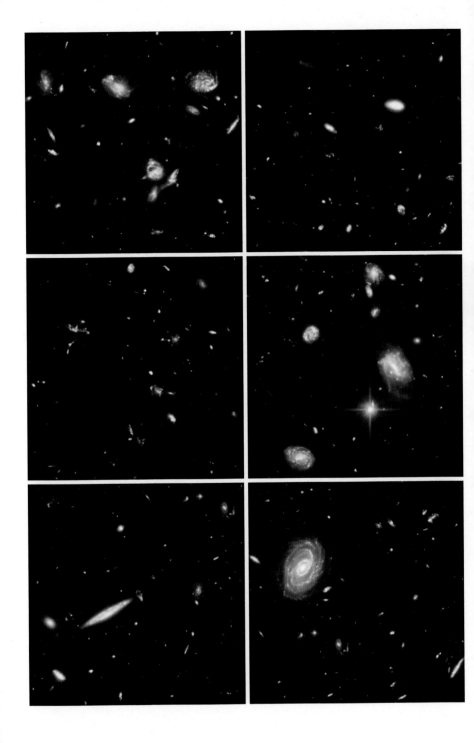

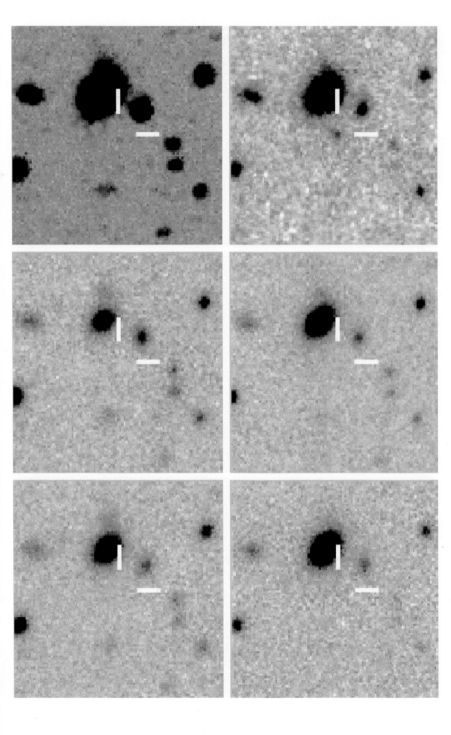

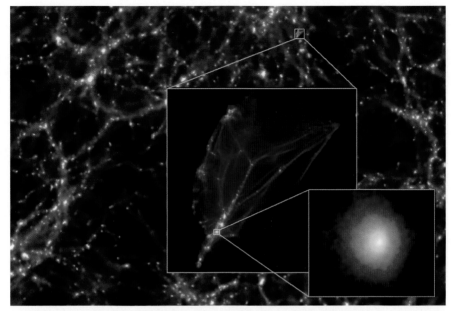

THE FIRST OBJECTS TO FORM
IN THE UNIVERSE

A supercomputer simulation shows how the first structures formed in the universe at a redshift of 26. The blue background depicts the filamentary distribution of matter 125 million years after the big bang across a span of 10,000 light-years. Brighter colors correspond to regions of higher density. The two inset regions are each expanded by a scale of one hundred. The final zoom (lower right) shows a single dark-matter halo that has the mass of Earth but the size of the solar system. Halos such as these are thought to be the seeds around which normal matter coalesced to form the first stars and galaxies. (J. Diemand, B. Moore, and J. Stadel, © Nature Publishing Group)

A GALACTIC WALLFLOWER

(OPPOSITE) The light from the most-distant galaxies has been stretched by expansion to longer, redder wavelengths. Astronomers can find these galaxies by comparing the same image field using different optical filters sensitive to these wavelengths. At a wavelength of 920 nanometers (top right panel), a tiny smudge appears at the lower right of the large galaxy in the field. At shorter wavelengths (lower four panels), however, the galaxy vanishes into the sky background. Neither is it apparent using a combination of different colors (top left panel.) The fact that the galaxy is visible only at this near-infrared wavelength indicates that it lies at a distance of more than 12 billion light-years. (J.-G. Cuby, O. Le Fèvre, H. McCracken, J.-C. Cuillandre, E. Magnier, and B. Meneux, "Discovery of a z=6.17 galaxy from CFHT and VLT observations," *Astronomy & Astrophysics* 405, no. 2 [July 11, 2003]: L19–L22)

What is the Reionization Era?
A Schematic Outline of the Cosmic History

Time since the
Big Bang (years)

~ 300 thousand

~ 500 million

~ 1 billion

~ 9 billion

~ 13 billion

◄─The Big Bang

The Universe filled
with ionized gas

◄─The Universe becomes
neutral and opaque

The Dark Ages start

Galaxies and Quasars
begin to form
The Reionization starts

The Cosmic Renaissance
The Dark Ages end

◄─Reionization complete,
the Universe becomes
transparent again

Galaxies evolve

The Solar System forms

Today: Astronomers
figure it all out!

A BRIEF PICTORIAL HISTORY
OF THE UNIVERSE

From the big bang to today. (Courtesy of S. G. Djorgovski and Digital Media Center, Caltech)

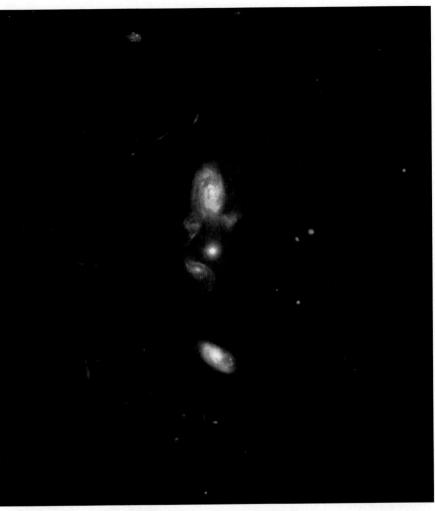

GALAXY MERGERS

The frequency of tidal interactions and galaxy mergers over different cosmic epochs is key for determining the star formation rates and galaxy assembly over the history of the universe. In this image, two bright foreground spirals (blue objects) have sideswiped each other and separated while in the distant background another pair (orange) come close enough to severely disrupt each other's disk. The blue colors in the spiral arms are young, hot stars that formed recently. (Image from the GEMS Collaboration)

A GRAVITATIONAL TELESCOPE

The gravitational well of the massive galaxy cluster Abell 2218 acts like a lens, bending and magnifying the light of fainter, more distant galaxies in the background. The reddish galaxy (shown in the unlabeled white outlines) is not only stretched around the cluster's center but is also multiply imaged. It has a redshift of about 7 and lies at a distance of nearly 13 billion light-years. The blue and orange arcs are lensed galaxies lying at intermediate redshifts. (Jean-Paul Kneib, Richard Ellis, and NASA/ESA)

He and Ellis were among the team members that bagged the red-shift 6.8 galaxy the previous year. Stark is a fresh-faced first-year graduate student who is observing at Keck for the first time. Because the success of his Ph.D. thesis is largely riding on the success of this observing run, Ellis has promoted him to team-leader status, something he has done many times before for other students.

This also explains why Stark is nervously monitoring the weather conditions on the summit. The skies over Mauna Kea are clear enough, but as attested by the three red warning lights on the computer's status screen, the humidity levels are still too high for the telescope dome to be opened. Humidity levels greater than 80 percent could result in condensation forming on the surface of the Keck mirror. This not only would diminish the overall image quality but could also leave condensation stains on the mirror, decreasing reflectivity.

"We really need about eight nights to show something," says Ellis, studying the red lights. "We have two nights now and another two in January. So we've got to apply for more time in October. If these nights are clouded out, then Dan has nothing to show for it. He has to say, 'Trust me, it's a great project. We have been very unfortunate.' If we lose January as well, he's going to start hitting the bottle." Santos chuckles at this, while Stark smiles grimly at his computer screen.

The objective for the evening is to look for gravitational lenses around massive galaxy clusters that might be magnifying more distant galaxies beyond. Their first target cluster of the evening, Abell 2218, is the one in which they found their redshift 6.8 galaxy. Astronomers have mapped the mass distribution around Abell 2218 and know exactly where the threadlike zone of highest magnification, called the "critical line," lies. Ellis shows me a Hubble image of the cluster upon which the critical line has been superimposed.

It follows a serpentine path around the cluster's periphery, much as Saddle Road meanders around Mauna Kea.

"There are several hundred lensing clusters in the sky," Ellis says, "but there are probably only twenty clusters in the sky that are known moderately well like Abell 2218. The area of the whole sky that has something in front of it that magnifies it by this amount is certainly smaller than the full moon."

"Less than one square degree out of forty-two thousand square degrees. Period," punctuates Santos.

In other words, they're looking through a keyhole in the sky for galaxies that are over a billion times fainter than the faintest objects detected by the unaided eye. Their only hope of catching a high-redshift galaxy lurking in this realm is to scan the critical line in a series of tight configurations, essentially creating a net in which to catch their galactic fish. This is where NIRSPEC enters the picture.

A modern-day spectrograph is a marvel to behold, but its power to analyze light is even more impressive. When a telescope takes a picture of a galaxy, it images the whole thing—all of its component stars, dust, and glowing gas. Spectroscopy, however, takes a different approach. The goal is to measure only a slice of a galaxy's light in order to learn something about the nature of the object at the point where the slice is made. It's like slicing into a steak cooking on the grill to see if it is overdone, medium, or rare.

The first thing that happens in a spectrograph like NIRSPEC is that light from a celestial object passes through a narrow slit. The slit Ellis and his team are using tonight is 42 arcseconds in length but a mere hairsbreadth—0.75 arcsecond—wide. Without restricting light to this narrow splinter, light from adjacent sources would pollute the purity of the spectrum, making it difficult to interpret. After passing through the slit, the light is then refocused onto a detector sensitive to a desired wavelength. In the case of NIRSPEC, its sensitivity ranges from 0.95 microns, the extreme red end of the visual spectrum, to 5.4 microns, which is well into the thermal infrared spectral region.

Positioning the slit angle precisely within the winding critical lines is itself an exacting task. Like Saddle Road, if you're not on it, you're wrecked. There is very little margin of error. It must be placed so that subsequent scans form a tiny grid pattern within the critical line. Each exposure lasts about ten minutes, enough to obtain a good, noise-free spectrum. Since this is Stark's project, he will be responsible for later "stacking" the sequences of exposures to see if they contain the signal of a high-redshift galaxy, a tiny dot of light within the otherwise smeared-out continuum of the foreground galaxy's starlight. The dot should fall at a discrete wavelength, that of Lyman-alpha hydrogen emission. As mentioned in Chapter 2, this line is normally seen in the ultraviolet, but at these extraordinary distances the line has been redshifted many times into the infrared spectral region where NIRSPEC is sensitive. If the dot falls around 1 micron, then Stark will know that it lies at a redshift of 7. If the dot appears at 1.14 microns, that would correspond to a redshift of 8; 1.37 microns would indicate a redshift of 10. Any signal detected by Stark would potentially break the existing record and be cause for celebration.

"What we're looking for," Ellis sums up, "is a blip of light from the magnified region [that is] a long, long way away."

But with the humidity levels still too high to open the domes, no one is doing anything. For the next two hours we munch on green grapes and sea-salt bagel chips and drink Diet Coke or coffee or both. Ellis and Stark excuse themselves to play a game of pool. We are joined by the lanky, easygoing Jim Lyke, a Keck support astronomer. He has written much of the NIRSPEC software the astronomers will be using tonight and is there to help with any difficulties that may arise. None are expected.

The lull in the action gives me the opportunity to pop next door to the Keck I control room to see what George Djorgovski and his Caltech collaborator Dave Thompson are up to. Djorgovski, who grew up in Belgrade (but in his words is a "born-again Californian"), is sitting in the control room's command chair, legs outstretched and fingers tented. Like Ellis and his team, they too are monitoring the little red status lights. Djorgovski is hoping tonight to, in his words, "push the frontiers of radio astronomy backward." He and Thompson want to explore the old 74-megahertz part of the radio spectrum to search for optical counterparts to radio sources. The frequency, which is lower than that of the FM broadcast band, is far below the usual frequencies (between 1 and 50 gigahertz) used for radio astronomy today. In fact, it was one of the first regions explored by radio astronomy pioneers Karl Jansky and Grote Reber in the 1930s and 1940s. Djorgovski and Thompson's idea is to select sources that are small on the sky, not just big radio galaxies, but ones that dominate around 74 megahertz and, if possible, image them in the infrared to see what they can see.

"There are three kinds of things we know of that can satisfy this criterion," Djorgovski explains. "Pulsars, about which I don't care. Big high-redshift radio galaxies, and also dying radio sources. Then again, [they might be] possibly something new and different, because nobody's ever looked this deeply at these frequencies."

Essentially, they cull the sources that don't meet this criterion and then take infrared images at Palomar Observatory of the sources that do. "This reveals a large fraction of them," says Djorgovski, "but there are some that we don't get. They're fainter in the infrared than even the most distant radio galaxies, which could be either because they're really far away or they're something else." He pauses. "Different."

Djorgovski, however, is accustomed to finding objects that are "different." In 1987, for example, he and his observing colleagues found the first known binary quasar, cataloged PKS 1145-071. This one object opened the door to a whole new world of possibilities as to how quasars may be "triggered" and how they maintain their prodigious energy output. Moreover, if quasars are rare objects, as Djorgovski reasoned at the time, then a binary quasar could raise timing constraints for the formation of structure in the early universe. Today there are only about fifteen known binary quasars, "plus or minus five," adds Djorgovski. Quasars are rare, but binary quasars are the rarest of the rare.

"It's not easy to grow something from a few tenths or even hundreds of solar masses into a billion solar masses in a few hundred million years," he says, referring to the supermassive black holes thought to lie at the hearts of quasars. "That requires very efficient creation. It's possible, but it's not so easy."

Although Djorgovski does not anticipate finding anything tonight as unexpected as a binary quasar, he doesn't necessarily dismiss the possibility altogether. The simplest explanation for the redness of the objects he and Thompson will be looking at is not that they're fundamentally different, just that they're really far away. "But if they're out beyond a redshift of 6," he says, "that would be really, really useful, because if you found a relatively strong radio source out there in the reionization era, then you could use that to probe the intergalactic hydrogen in a whole different way. It would be a different physical probe. It's a few of those that we really want to nail tonight . . . if everything dries out."

Back in the Keck II control room, Ellis and Santos are bantering about the universe in general, and the early universe in particular, all the while watching the stubborn red lights. Stark is parked at the main computer nearest the teleconference monitor going over the observing plan, probably for the hundredth time. Just before eleven o'clock, the red lights change from red to amber, then finally to green. Ellis is in the middle of a comment when he suddenly exclaims, "Hey, a green! Look!" Twenty minutes later, at 11:15, Steven Magee, the Keck II telescope controller, announces that the dome is being opened. After a few minutes Magee looks into the teleconference camera and asks nobody in particular, "Where do you want to go?"

Ellis, Santos, and Stark agree on their first target area around Abell 2218, and within seconds the great telescope begins to direct its aim toward that small region of sky. The telescope status screen shows two sets of sky coordinates. On the left are the fixed target sky coordinates, while those on the right are rapidly changing, showing the progress of the telescope as it slews slowly to the target area. Suddenly a mechanical voice pipes up over the monitor. *"ACS centering is not normal. ACS is not settled."* The message repeats. ACS, Lyke leans over to explain to me, stands for active control system. It's responsible for maintaining the optical figure of the primary mirror's thirty-six segments under the changing influences of gravity and temperature as the telescope guides. The computer's verbal warnings, however, can at present be ignored, he says. "The ACS is not meant to maintain a perfect shape while slewing, only when tracking objects."

A single star drifts onto the computer screen. The group members are quieter now, murmuring among themselves against a background of whirring hard drives, squeaky chairs, and the disembodied litany *"ACS centering is not normal. ACS is not settled . . . ACS is not settled . . . ACS is not settled."*

"Okay, we're getting 0.6 arcsecond seeing," Magee tells them.

"So that's very good seeing," intones Ellis. He turns to Stark and says brightly, "Maestro?"

Stark gives the command to Magee in the telescope control room, and Keck II begins its first exposure of the evening. Within minutes the central computer screen displays a slightly tilted rectangular picture of white and dark lines. It looks like a picket fence that has had some of the pickets randomly knocked out. This is the spectrum taken across the galaxy cluster's critical line. No apparent dot appears in the spectrum, but that is to be expected. If one is there, it will take many more combined exposures to raise it to the surface.

For over an hour the astronomers systematically step their way across the critical line with more exposures. By 1:45 a.m., however, the team is getting nervous. Abell 2218 is now very high in the sky—far higher than the 75-degree altitude that is normally Keck's limit.

"Jim has just reminded me that we're about 75 degrees [altitude] already," Ellis warns, "so we don't have that much more [time] on the subject." The Keck has an altazimuth mount. Like a great circus cannon, it can move up and down in altitude and swivel horizontally in azimuth. However, if the telescope tracks through the zenith, the straight-up point in the sky, the NIRSPEC slit will pivot around with the telescope away from the desired angle across the critical line. This can wreak havoc with the NIR-SPEC software program, which wants to keep the slit angle fixed.

"The instrument is talking to the telescope—ferociously," says Ellis. "It might get its knickers in a twist. Above 80 degrees it starts to get dicey."

Minutes later, as Stark watches quietly, Santos begins calling out elevation numbers: 82 degrees, then 86. By 87 degrees—just short of vertical—warning messages begin to flash on the screen. The exposure is nearing its end when the server crashes. A ten-minute exposure of the sky is lost. Lyke jumps in to try and get the system up and running, but just ten minutes later the computer crashes again. It takes nearly a half-hour before Lyke and Magee can reboot the system.

At 1:55 a.m. the problem appears to be resolved, and the tension that had been mounting in the control room dissipates. Ellis announces with relief, "We're off," and the astronomers continue their spectroscopic scans across the critical line of Abell 2218. Despite the earlier glitches, they complete a total of four and a half hours of exposures for the evening. The following night they meet with even greater success. Good weather, excellent seeing conditions, and no computer crashes enable them to scan all night. Stark must now take all this hard-earned data and begin the months-long task of reducing it. With any luck a redshift 8 or 9 galaxy will pop up, advancing science and clinching his Ph.D. thesis.

Despite its extant successes, sifting the high-redshift universe with gravitational lenses is fraught with uncertainty. "It's a very risky way of surveying," Ellis says. "You really hope that these [galaxies] are common on the sky. Otherwise, you could scan these tiny areas and see absolutely nothing."

The ultimate goal of these deep surveys is to formulate a more precise timeline for cosmic reionization. How long would it take for the stars in high-redshift galaxies to light up the universe? The answer depends on their abundance, mass, and luminosity at redshifts greater than 6. Right now that is an unknown, but observers like Ellis and his colleagues are trying to find out. Their long-range plan calls for them to observe up to twelve galaxy clusters over the next two years with enough sensitivity to see the kinds of objects that might have caused reionization. These will likely not be the first crop of metal-free stars, says Ellis, because they form in isolation, and although they are spectacularly luminous, they are also short-lived.

"I suspect that most of the energy needed to break up the hydrogen and ionize the universe is coming from sustained objects that are assemblies of stellar systems, stellar populations. Take our object at redshift 7. It's a few hundred million solar masses. It's a monster; it couldn't possibly be a single object. It's got to be a stellar system of some form. We know how big it is as well. It's been shining for a long time, so if the abundance of these objects is reasonably typical of the fact that we found one just by looking through this tiny area, then we're not that far off from the number of photons needed to reionize the universe. . . . We're searching for what is out there, and how common these objects are. It's like setting sail across an unknown ocean and not knowing what you're going to find."

Following our two-night observing run, Alex and I reward ourselves by flying to Honolulu to visit a friend and to finally do what most tourists are supposed to do in Hawaii—go to the beach and sip mai tais, although not necessarily in that order. It is a short but sweet stay. Unfortunately, the trade winds are flagging and not even the memory of the spectral blue waters of Hanauma Bay can keep us cool in our friend's unair-conditioned apartment at night. I miss the verdant hillsides, cool breezes, and mild temperatures of Waimea, but mostly I miss the marvel of Mauna Kea. One need go there only once to fall in love with the White Mountain.

Two days later, as we are winging our way back to the mainland on a late afternoon flight, I glance out the window and look upon a magnificent vista: a dark hump-shaped mountain wreathed in clouds catching the long-slanted golden rays of the setting Sun. It casts a vaulted, deep-purple shadow eastward across the pewter-colored Pacific Ocean. Only when I turn my attention to the summit itself and see several white dots clustered there do I realize I am looking at Mauna Kea. The two dots on the left would be the domes of the Canada-France-Hawaii and Gemini telescopes. The ones on the right, appearing nearly merged into one, are the Keck domes. No doubt astronomers and technical support staff there and in Waimea are preparing for another night of sounding the universe. The sky is certainly clear enough. Once darkness falls, and if the humidity is low and the wind calm, the domes will open one by one and the telescopes will slew to their assigned targets— dusty stars, glowing nebulae, and far, far away galaxies. What more will astronomers find tonight? Just tonight? How much more *is* there of the universe to find? Certainly more than can ever be found or understood.

I follow the receding summit of Mauna Kea until it passes from view beneath the wing. I can still see the mountain's broad shadow below me, bigger than the mountain itself, its apex rising above the greater shadow of the Earth's limb.

8

PECULIAR UNIVERSE

Not all galaxies fit the schematic idealization of the Hubble sequence of nebular forms. In fact, when looked at closely enough, every galaxy is peculiar. Appreciation of these peculiarities is important in order to build a realistic picture of what galaxies are really like.

—Halton Arp, *Atlas of Peculiar Galaxies*, 1966

There is a well-known spiral galaxy called the Whirlpool that in my telescope looks exactly that: a feathery swirl of diaphanous light wrapped like a watch spring around a bright central sprocket. The symmetry is broken only by an outer arm extending toward a smaller galaxy that, millions of years ago, swept close enough to give the disk a gravitational stir. The wayward appendage is an added flourish to an already sublime object.

The Whirlpool is exceptional in that it is one of the few galaxies bright enough to actually look like a spiral galaxy in a backyard telescope, without the aid of long-exposure photography or electronic imaging. Most galaxies are just too faint for their structures to be appreciated visually. Over a century's worth of observations with large, professionally equipped telescopes, however, has allowed astronomers to catalog tens of thousands of galaxies that, more or less, are analogous in form to pinwheels, propellers, and variously shaped spheroids. But as we have seen, within the last dozen years deep surveys have turned up myriad weird-looking galaxies at distances greater than 11 billion light-years. If the diversity of galactic shapes makes them compelling studies, it also inspires a very good question: Why do galaxies look the way they do?

Historically, astronomers have addressed this question by systematically sorting galaxies by their fundamental shapes. The hope was, and is, that the arrayed types may allow astronomers to draw conclusions about galactic properties and evolution in the way that a family tree tracks the developmental history of bloodline relations. This taxonomic approach (borrowing from the branch of biology that deals with the form and structure of plants and animals) is known as galactic morphology. Over the past three-quarters of a century, various classification schemes have been proposed, but most stem from Edwin Hubble's original 1925 system, which arranged galaxies into a two-pronged "tuning fork" scheme. The tuning fork's handle consisted of orblike elliptical galaxies, with spirals arrayed along one prong and barred spirals (propeller-shaped galaxies) along the other. Each galaxy was assigned an alphanumeric code that signified its location within the sequence.

Though Hubble's scheme has been augmented many times since—most notably by his protégé, Allan Sandage—its modern incarnation still adheres to a straightforward linear sequence. Four broad galactic families are recognized: spirals, ellipticals, lenticulars (tapered, lens-shaped systems), and irregulars (sedate, patchy conglomerations of stars without any definite form or rotation).

Tightly wound spirals with prominent central bulges are classed Sa and Sb, while the loosely bound ones with stellarlike nuclei are Sc and Sd. Barred spirals are organized according to how tightly their outermost arms coil about their central regions. Arms that are tight and nearly ringlike are ranked SBa and SBb, while those that are very loose or broken are SBc and SBd.

Elliptical galaxies are ranked by how round or flattened they appear. The most spherical ellipticals are labeled E0, while those that are highly flattened are E7. The lenticulars—S0—act as a kind of bridge between the E7 ellipticals and spiral disk galaxies, although the two types have different evolutionary tracks. Lenticulars generally contain older stars, while spirals are bedecked with strands of hot young stars.

All other galaxies are placed in the irregular category, including the small blue galaxies that lack organized structure—called dwarf irregulars—as well as the Milky Way's two neighboring systems, the Large and Small Magellanic Clouds. The Small Magellanic Cloud is considered a normal irregular and classed Irr, while the Large Magellanic Cloud, which exhibits a prominent bar, is considered an SBm (for Magellanic barred-spiral, yet another subcategory). Irregulars typically contain greater amounts of interstellar gas than other galaxies, as well as a larger fraction of young, low-

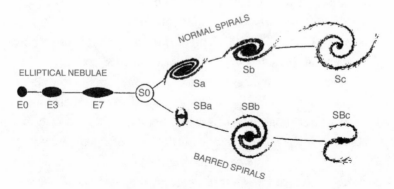

The Hubble tuning fork. (Courtesy Yale University)

metal stars. Some irregulars exhibit strong star-formation activity as well, but their low-metal abundances indicate that they have only recently "turned on."

One of the more functional aspects of the Hubble scheme is that it clearly shows how galaxies are physically similar as well as dissimilar. Even irregular galaxies are recognizable as a class, though they do not share the same shapes. Such distinctions provide a kind of mental shorthand for astronomers. Evoking an Sc-type galaxy, for example, paints an instant image, complete with physical properties and dynamics.

But how useful is that, really? It seemed to be highly promising in the 1930s and 1940s, when it was thought that galactic shapes and ages might somehow be related—with ellipticals, or "early" galaxies, evolving into spirals, or "late" galaxies. But that notion fell by the wayside long ago (although astronomers still refer to ellipticals and spirals as early and late galaxies). Today factors such as galactic environment and initial mass and angular momentum are considered responsible for sculpting galaxies. There *is* an evolutionary path, to be sure, but like a wild river it meanders.

There is another problem not adroitly dealt with by most of the conventional galaxy classification schemes of the mid-twentieth century, and that is that while most galaxies line up neatly behind galactic archetypes, not all of them do—particularly, as we have seen, those at the detectable limits of the universe.

This was not so much of an issue in the mid-twentieth century. Back then astronomers had a kind of junk-drawer designation for the few galaxies that defied classification: they were simply called "peculiar." Like the human-beast experiments gone wrong in H. G. Wells's *The Island of Dr. Moreau*, peculiar galaxies appeared to be the results of some sort of experiment gone wrong. The most unusual examples appeared to be composites of different types of galaxies, a battle among mongrel spirals or between ellipticals and spirals, with no dominant form winning out. Many exhibited appendages and filaments drawn out more than twice the length of the galaxy and into intergalactic space, leading nowhere. Fortunately, there weren't a lot of peculiar galaxies—less than 2 percent of all known galaxies—so most galaxy morphologists at the time felt they could afford to shelve these outcasts, at least until the dynamics of normal galaxies were better understood.

From the early to mid-1960s one person—Mount Wilson astronomer Halton C. Arp—was largely responsible for studying peculiar galaxies, simply because no one else wanted to have anything to do with them. In 1966 Arp compiled and published an atlas of 338 peculiar galactic systems. His goal was not so much to try to classify them as it was to draw attention to these cosmological stepchildren. His research gave him a level of insight that proved particularly prescient. In the introduction to his *Atlas of Peculiar Galaxies*, Arp stated that *all* galaxies, when looked at closely enough, are peculiar. This notion, although acknowledged up to a point by most astronomers, was the equivalent of a mission statement. Galactic peculiarities, Arp believed, represented perturbations and deformations perpetrated on otherwise normal galaxies. Analysis of these would help astronomers better understand the "special events and reactions that occur in [normal] galaxies."

Throughout the 1960s and 1970s galaxies were classified by their appearance in visual (or blue) light. In subsequent decades multiwavelength observations revealed additional structures and energy sources beyond what could be detected within the narrow optical band. The beautiful cochlear appearance of many normal spiral galaxies in visible light gave way to patchy chains of superhot gas or smoldering star-forming regions in the ultraviolet or in-

frared, while tiny starlike nuclei swelled into blazing protuber-
ances when viewed at ultraviolet and X-ray wavelengths. How
could such features be reconciled with the otherwise symmetrical
appearance of a "normal" galaxy?

By the mid-1990s, when galaxies beyond about 5 billion light-
years began to emerge in deep images, it became clear that indeed
Arp had been correct—*all* galaxies are, in some sense, peculiar. In
fact, today a galactic morphologist would be hard-pressed to point
to any galaxy, even our own, and deem it "normal."

The knowledge base of galactic research is today so broad and
bounteous that new methods of galaxy classification are being
sought. Astronomers openly admit that the time-honored Hubble scheme has outlived its usefulness. "The traditional Hubble
scheme was instructive for a while, but I think we just have to
grow up and accept the fact that galaxies are not as simple, not as
one-dimensional, as we may have thought," says Barry Madore, a
galaxy specialist with the Carnegie Observatories and collaborator
with Arp on a second momentous work, *Catalogue of Southern
Peculiar Galaxies and Associations*. "It makes it very hard to talk
about things that have seventeen dimensions to them."

Madore would like to move away from conceptually laden classification schemes toward a system where galaxies are described in
a context-free way that doesn't predefine what people think a
galaxy is doing. "Most spiral galaxies in the Hubble Atlas are two-
armed structures, but what makes two more interesting than one,
or three, or fifty?" he asks. "You won't find any indication of the
numbers of spiral arms in the Hubble classification, but it may well
be that through interactions one only gets one-armed spirals, that
bars or other instabilities are generally found in two-armed spirals,
and three-armed spirals may be some other form of instability."

Madore is hopeful that some sort of workable classification system may one day be in place, but he is nonetheless aware of the potential pitfalls in how such a system might be applied not only to the kinds of peculiar galaxies seen in the deep field surveys but to *any* galaxy, near or far.

"Chaos theory shows there are limits," he says. "Who goes in and classifies what turbulent eddies look like? No physicist in their right mind would classify that sort of thing. And maybe that's all we're doing. We have this frozen picture in time of things that don't look like what they used to look like, probably won't look the same in the future, and don't even look like what they *do* look like when you look at them in another wavelength! It may well be that every galaxy's structure is so contingent on the last event or environment that it found itself in, that what you see now in this fine little slice in time and this other tiny slice in wavelength have nothing to do with its global history—the last billion or ten billion years. To reduce a whole galaxy down to two letters and a number seems to trivialize the whole thing."

For Simon Driver, a galaxy specialist at Australia's Mount Stromlo Observatory, the Hubble tuning fork is just butterfly counting. "The name of the game is to find something to replace it with," he says. Some of the new structural approaches factor in such properties as bulge-to-disk ratios, bulge profiling (how concentrated a galaxy's bulge appears), degree of asymmetry, and spectral classification. Many of these alternatives strive to connect features back to the Hubble tuning fork. "Personally, I think that's the wrong way to go," says Driver. "Galaxies comprise more than one component. In the case of our galaxy, we now have two bulges, two disks, and a halo—five components."

Any approach to classifying galaxies can be an exercise in tortured semantics and multilayered abstraction, but perhaps a starting point would be to agree that if no galaxy is normal, then there must be various levels of not-normalness. Put the other way around, if we agree that all galaxies are peculiar, we must also make the Orwellian concession that some galaxies are more peculiar than others. Peculiarities, then, can range from "nearly normal" to "abnormal" to "off the charts." Moreover, peculiarities can be rated as a function of wavelength. Only 1 or 2 percent of nearby galaxies are oddballs in visual light, but practically all of them are peculiar to some extent when observed in other wavelengths.

Of course, by far the greatest number of abnormal galaxies appear in the Hubble Deep and Ultra Deep Fields, in which ultraviolet light shifted into the near infrared provides the dominant form of luminosity. The majority of these galaxies lie in a universe of long ago and far away—at redshifts between 2 and 6—implying a clear relationship between distance and galactic asymmetry. Christopher Conselice has made it his mission to try to understand the history and cause of this relationship.

Conselice, a Caltech astronomer and galaxy morphologist, argues convincingly that the shapes of galaxies change with time. This concept may seem to be a no-brainer when one is comparing photos of nearby galaxies with those in the Hubble Ultra Deep Field, many of which lie beyond a redshift of 1, but determining the history and cause of this relationship is the real challenge. In fact, says Conselice, it may be the missing link in our understanding of how galaxies are formed.

Whereas most people see a riot of galaxy shapes in images like the Hubble Ultra Deep Field, Conselice discerns an overall pattern that helps him understand why these galaxies appear the way they do. "Galaxies at redshifts greater than 2 are fundamentally different than those today," he says. "They are of lower mass and have distorted morphologies that are the result of galaxy mergers and interactions. The reason for [the mergers] is that the density of the universe is higher at high redshift, so you would expect to have more galaxy mergers, which is indeed the case."

Hence not only have galaxies changed in shape with redshift, so has the frequency of mergers and collisions. This explains why, in the nearby universe, the peculiar galaxies ascribed to interactions are few and far between. Mergers are short-lived events, which is also a contributing factor in their scarcity today. The further back in time you look, however, the more mergers you see. The deep and medium-deep surveys have afforded astronomers a panoramic view of galaxy mergers across various redshift ranges. From these observations they can mathematically model how much mass a galaxy will acquire over time as a result of these mergers. A large fraction of the bright modern galaxies are either ellipticals or spirals with masses exceeding several hundred billion suns. But mergers of galaxies beyond a redshift of 2 don't produce galaxies nearly that size. "The important thing to realize is that the most massive galaxies at redshifts of about 3 or so are only as massive as the bulge of our galaxy," says Conselice, "and our galaxy is just of average mass."

Mergers, collisions, and "sideswipes" also explain why these galaxies look like so many thronging amoebas on a microscope slide. Gravitational interactions stir up a galaxy's gas, altering its density distribution, which in turn triggers disproportionate star-forming regions.

"Star formation occurs when gas becomes dense, and this results in new stars being formed in clumpylike regions," says Conselice. "You can see this in our own galaxy today and in other nearby galaxies. When two galaxies merge, they become distorted and asymmetric. When they finally finish merging, they relax and become smooth, like an elliptical. This is what is going on with these [high-redshift] objects. The upshot is that we can learn about how galaxies are forming through their structures, or how the stellar light is distributed."

Conselice admits that the idea of using the structures of galaxies as a way of understanding their current and past formation histories is new and perhaps a little controversial. However, he thinks there is more than enough evidence to suggest that some galactic structures correlate strongly to physical properties. Some of these properties, he points out, are the current rate of star formation, the total stellar mass, the galactic radius, the mass of the central black hole, and the merging features. The light concentration of an evolved galaxy, for example, correlates with its luminosity, mass, and size, while the central light concentration also corresponds to the mass of central black holes. A galaxy's "clumpiness" can be associated with the amount and location of ongoing star formation, while its degree of asymmetry points to its dynamic history. Merging galaxies tend to be not only more asymmetric than nonmerging ones but also more luminous and massive. In fact, roughly half of all the bright or massive galaxies Conselice has studied in the Hubble Deep Field appear to be undergoing major mergers, while less than a quarter of the fainter and less-massive galaxies are experiencing mergers.

Simon Driver, whose primary research thrust is surveying nearby galaxy populations and tracking their evolutionary paths, has his reservations about Conselice's approach. "The main problem is that galaxies basically form a puddle," he says, "and it's somewhat arbitrary as to where you cut it. The main worry is that systematic biases may shift the puddle around, so rather than detecting clear trends in evolution, one detects trends in the biases."

Driver favors a "bulge-to-disk decomposition" strategy in which a two-component model is fit to a galaxy's image. One component represents the central bulge, while the other represents the surrounding disk. Some galaxies are all bulge (ellipticals), others contain both bulge and disk components (lenticulars and spirals), and still others are all disk (irregulars and most dwarf systems). The bulges and disks are known to follow different dynamics, says Driver: "Bulges are pressure-supported and the stars move in all directions with a range of speeds, whereas disks are rotationally supported, and the gas and stars all orbit in the same direction in a flattened plane. We also know that bulges and disks have different stellar populations, metallicities, gas, and dust contents. Hence it seems logical to reduce, or decompose, the problem of galaxies to one of bulges and disks."

A galaxy's bulge-to-disk ratio is also dependent on the environment in which it forms and evolves. Galaxies located in the cores of rich clusters are typically bulge-dominated, whereas those located in the lower-density regions are disk-dominated. This suggests that the first galaxies didn't form in isolation but must have been involved in interactions with other protogalaxies early in their evolution.

One of the bulge-to-disk concerns is whether it can be applied to the messy galaxies seen at higher redshifts. Driver believes it can, to some extent. "Some people argue that, yes, most galaxies in the Hubble Deep Field that can be modeled with a bulge-disk decomposition will include 80 percent of the light, even for highly disturbed things. But it becomes harder as galaxies become more and more irregular in appearance. Is this because of more dust, the fact we're sampling ultraviolet light, the high fraction of young star bursts, or because they're fundamentally different and don't adhere to the bulge-disk categories? More likely we need to move to the near infrared, where galaxy images are smoother, and to the mid and far infrared for the high-redshift galaxies. This is one of the reasons why we're so keen on the James Webb Space Telescope. It should provide this kind of data."

Although many challenges remain in sorting out viable classification methods for distant galaxies, the early results look promising. "This is still very much in progress," says Conselice, "but we see the outer parts of distant galaxies, and we can study in more depth the structures of higher-redshift systems. What we've found so far is that the highest-redshift galaxies look like a merger population, but with lower masses. Also, it appears that the star-formation rate stays relatively high all the way out to the highest-redshift galaxies we can now see."

Recently, astronomers have taken a cubist approach in trying to fathom what they're seeing in the various deep fields. They begin by asking themselves, "What if we took this odd object and rotated it 90 degrees? What would it look like?" Much of galaxy morphology depends on a fortuitous viewing angle. From Earth, spiral galaxies seen edge-on can look like concave lenses or flying saucers. When seen face-on, they may exhibit the grand spiral design we all admire in coffee-table books and planetarium movies and such. But for the high-redshift universe, there are no obvious analogs. Nevertheless, the cubist approach has led to some interesting observational insights that may have a direct bearing on accepted theories about how galaxies coalesce into the ones we see in the modern universe.

In 1995 a team of astronomers led by Lennox Cowie from the University of Hawaii announced the discovery of a new morphological class of galaxy in very deep Hubble images. These galaxies appeared to be narrow, linear structures with superposed bright knots. Cowie's group called them "chain" galaxies. Some appeared nearly streaklike, while others resembled a sequence of discrete blobjects, like a more distant version of the Shoemaker-Levy 9 comet that broke into a train of pieces as it fell into Jupiter in July 1994. The fact that the galaxies were very faint implied that they were extremely distant objects, but the nature of their segmented structure was a mystery.

Then in 2004 astronomers Debra and Bruce Elmegreen and Amelia Hirst reported in the *Astrophysical Journal* their discovery of what they interpreted as the face-on counterparts to Cowie's chain galaxies. They called them "clump clusters." An archetypal clump cluster looked like an irregularly shaped oatmeal-raisin cookie, with the raisins being analogous to the clumps. The Elmegreens and Hirst based their conclusions on the fact that the brightest clumps in the clump clusters were similar in color and apparent brightness to those in chain galaxies. What was significant, however, was the fact that neither chain galaxies nor clump clusters exhibited any prominent knots that would qualify as a bulge; nor did they seem to have a well-defined disk. The fact that neither chain galaxies nor clump clusters possess an obvious bulge component implies that perhaps the formation of a bulge occurs after the disk forms—not before, as has been generally assumed.

Shapeless though primordial galaxies may appear, their components nonetheless provide hints of their formation history. Taking this idea a step further, these galaxies may also provide pertinent clues as to how the early universe evolved. As the WMAP data and high-redshift quasar surveys show, by a redshift of 6 (950 million years after the big bang), the universe had produced enough luminous sources to have completely burned off the neutral hydrogen fog, making the universe transparent to photons. What kind of sources are we talking about? The candidates range from massive galaxies to quasars and mini-quasars to galaxies containing massive stars. Sometimes researchers' conclusions stand poles apart, with one group claiming that the combined ultraviolet light of galaxies at redshift 6 is insufficient to reionize the universe, while another argues that it is sufficient.

One way to go about determining the effectiveness of potential reionizing sources is to perform a kind of census of objects that live in that epoch. First astronomers count how many galaxies there are of different brightnesses in a standard volume of space at the critical redshift interval between 5 and 7—a period that spans only 420 million years or so. From the galaxy counts you can then make assumptions about how numerous they were during that period of time and whether they possessed the radiant power to reionize the universe by a redshift of 6.

In terms of population, giant galaxies were few in number and could not have completed reionizing the neutral hydrogen medium on their own. Quasars, too, were scarce and hence make unlikely primary sources. But by far the most abundant type of galaxy that existed at the cosmic dawn was the humble dwarf galaxy. These systems are intrinsically faint and small. At redshifts of 5.5 and 6, their bantam sizes transect an angle on the sky of just 0.1 to 0.2 arcseconds; thus they are only about 6,300 to 13,000 light-years across. For comparison, just the bulge of our Milky Way is a little over 9,000 light-years in diameter; so these galaxies are indeed dwarfish in size and luminosity. But could these faint, feeble systems effectively contribute to ending reionization? Astronomers Haojing Yan, of Caltech's Spitzer Science Center, and Rogier Windhorst, of Arizona State University, decided to find out.

First they sifted the Hubble Ultra Deep Field for visible dwarf galaxies in the redshift 6 range, finding 108 candidates. From these counts they extrapolated the fraction of fainter galaxies that were not apparent in the HUDF image and discovered that the number of dwarf galaxies rises swiftly as brightness levels fall. Yan and

Windhorst concluded that the entire ensemble of dwarf galaxies could collectively generate enough ultraviolet photons to consummate reionization by a redshift of 6. This finding, says Windhorst, helps give them some idea of the birth date of these dwarf galaxies: "We think they started to form in significant numbers at redshifts 7.5 to 8." That's roughly 300 million years before their redshift 6 showing.

Obviously, there is much to be learned from these budding galaxies, whether they are assayed by their bulge-to-disk ratios, their asymmetries, their compactness, or their colors. The bottom line is that they are all exceedingly peculiar and clearly indicate, in an old-fogy kind of way, that galaxies back then certainly aren't like galaxies these days. Today they are more staid, more massive, and structurally closer to normal galaxies on large scales. At some point between the then-and-now universe, these peculiar galactic striplings settled down. But when? One survey may have answered that question.

One of the more prominent successes of the Gemini telescope is how it has been used to probe the universe when it was between roughly 3 and 6 billion years old, corresponding to redshifts between 1 and 2. Galaxies existing at this time are particularly difficult to study because most are very faint. The brighter ones are visible only because they are undergoing intense star formation. Spectra have been taken of the brighter, energetic galaxies, of course, but they are few and far between and don't represent the vast majority of galaxies that have evolved to this epoch. Astronomers dub this unexplored part of the universe the Redshift Desert.

Studying galaxies in the Redshift Desert is important for two reasons. First, it was during this epoch that the galaxies began assembling themselves in earnest. Second, it was also the period in which galaxies assumed more "normal" forms, like the ones we see pronged onto the Hubble tuning fork. But it was also here that precious galactic light so vital to astronomers dried up. In truth, the light is still there, but it was thwarted in its journey to Earth's surface. The culprit is the glow from our own sky. Even from the darkest locations on Earth, the sky emits a pallid radiance, which is the result of emission from the hydroxyl molecule present in the upper atmosphere. Visual light from space is obscured by this skyglow, particularly if the source is intrinsically faint. Further aggravating the situation is that ultraviolet emission evaporates in the Redshift Desert, since at these intermediate lookback times it is less redshifted toward longer wavelengths by cosmic expansion.

Researchers using the Gemini telescopes for a project called the Gemini Deep Deep Survey (GDDS) have been able to venture far into the Redshift Desert by employing a telescope-pointing technique called "nod and shuffle," which filters out skyglow. With this technique Gemini astronomers were able to make exposures thirty to fifty hours long, whereas in the past they were limited to five-to-eight-hour exposures.

Here's how you do the nod and shuffle. After you obtain a spectrum of an object, the electrical charge that was built up by the image is moved, or "shuffled," to a storage buffer on the charge-coupled device (CCD). Next, the telescope's position is slightly offset, or "nodded," and a spectrum is made of just the sky. The charges shuffled from the first exposure are then shuffled back to the original position while the charges from the sky exposure are shuffled to the storage buffer. The telescope is then nodded again back to its original position, where another spectrum of the object is taken. The process repeats until sufficient light is collected. The final step is to take the two spectra produced and subtract them from each other, removing the unwanted atmospheric component. The resulting spectra are then combined into a final spectrum.

Using this technique to eliminate skyglow, Roberto Abraham of the University of Toronto and a team of American, Canadian, and British astronomers gathered spectra of three hundred galaxies over the course of a year, enabling them to obtain a large sample of spectra from the fainter, more established galaxies. To their surprise, they found that a large fraction of the stars in galaxies in the Redshift Desert were already in place when the universe was quite young.

"The theoreticians will definitely have something to gnaw on," Abraham said to reporters at the January 2004 meeting of the American Astronomical Society. Months later, in a telephone interview, Abraham and team were still speculating about what the implications might be, while finishing up additional science papers on their work.

"It was a big surprise that these galaxies are out there and they're full of old stars," says Abraham. "But it's not clear that you can't explain them within a hierarchical galaxy-formation picture, provided you come up with a means of making star formation occur surprisingly early in the universe and then just shutting it off. You have a big episode where you make the galaxy at redshift 3, 4, or 5, but it only forms all its stars for about 200 million years and then just shuts off. At the moment we have to come up with a mechanism for doing that, and nobody's come up with that so far."

It sounds like a job for the theorists, but Abraham doesn't think it is. "I used to think that was the right thing to do, but I've now concluded that the right way to do this is totally empirically. The theorists are stuck because in order to make real progress, I figure those guys need to tell me how stars form in galaxies, not just in our own galaxy. I figure the right way forward here is to build better spectrographs with bigger telescopes and just nail the problem observationally."

One of the paradoxes posed by observations of galaxies in the early universe is the apparent predominance of elliptical galaxies, which have somehow fleetingly transformed most of their gas into stars at high redshifts. Elliptical galaxies are usually considered to be morphologically advanced systems consisting predominantly of old red stars. Seeing these systems established in a universe that is around 3 billion years old, therefore, is a bit unsettling. Dense regions of huge dark matter halos are required to create ellipticals so soon. That, however, would challenge the favored hierarchical model, which holds that massive galaxies are built up by the many mergers of smaller galaxies.

"If you smash two disks together, it's going to look like an elliptical galaxy," says Abraham. "By the same token, you have to wonder whether that's the only way to make an elliptical galaxy. There's abundant evidence from other parts of my argument that indicate that the stellar populations of ellipticals are old. So how do you get past this apparent paradox? My intuition now is that how galaxies form is a strong function of their mass. We see this with the GDDS, too, where the most massive galaxies form first and the less massive galaxies form a little later. I think it's imperative to think of this in terms of the mass of a galaxy. The massive galaxies probably formed at pretty high redshift through some mechanism that I don't understand and I don't think anyone understands. The less-massive ellipticals—this is just me guessing—probably originated through merging disks. I would presume that different mechanisms were important at different times of the universe."

Abraham refuses to wait for the theorists. More redshifts need to be measured for galaxies at greater distances, he says, and then a census made of the number of big disk galaxies.

"I don't think that the models are by any means useless," he concedes. "They're actually very useful for suggesting what sort of observations to get. But I think the observations condition the models. Right now we're in a phase where the observational component is really in the ascendancy. In order to make real progress, I think that the models have to come up with a mechanism describing how you form stars and dark matter halos. Star formation is such a messy thing. When we have a real believable model for star formation, then I think the cycle will turn and theory will be driving the observation more than the observation will be driving the theory."

There can be no doubt that the galaxy game is afoot for observers. Edwin Hubble assumed that galaxy shapes could be placed into an evolutionary sequence. He was right, but about three-quarters of a century too early. Astronomers' newfound wealth of galaxies has also posed greater challenges, however, namely: How does one effectively connect the evolutionary dots of galaxies in the modern universe to the ones that existed before Earth even had a past—back to a time, in fact, when the Milky Way was so much drifting gas and dust twinkling in God's eye? The best way, it seems, is to conduct surveys that rake in large numbers of galaxies at various redshift distances, thus sampling them at different stages of evolution. The hope is that one galaxy can then be linked to another without making the wrong connection.

"It's like taking a picture of a forest at various distances from you," says astronomer Eric Bell, with the Max Planck Institute for Astronomy in Heidelberg, Germany. "You may see a flower, a bush, a small tree, a big tree, and then a log. You may connect these to an evolutionary sequence: a flower becomes a shrub, becomes a small tree, becomes a big tree, becomes a log. And while parts of that are perfectly right, parts of that are complete crap."

It's virtually impossible to determine from observations alone how something as complex as a galaxy evolves over time when the time frame is inaccessible to the observer. Compared to human life spans, galaxies don't change quickly. To an immortal astronomer, changes at stellar scales may be observed in other galaxies over a few hundred million years, but it would take closer to a billion years before any significant large-scale structural change could be observed. Computer modeling might help astronomers make fairly accurate predictions, but what about unpredictable changes in a galaxy's environment or star-formation rate? How can a central black hole's behavior or a spate of supernovae be conclusively factored into the scheme? If all we really have, as Barry Madore says, is a little "slice of time," what can that convey about future or past slices of any galaxy?

Fortunately, we do have one thing that may be of use in understanding how galaxies have evolved since the universe began. We have our own galaxy, the Milky Way, as well as a variety of nearby galaxies, each with a past that has inexorably unfolded from those same beguiling objects that flourished 13 billion years ago. Like photographs of our parents, grandparents, and great-grandparents taken when they were children, these young galaxies glimmer from a time that precedes our own existence but to which we are nonetheless linked. By bringing them to light, astronomers have revealed a great deal of our cosmological heritage. Those galaxies aren't simply related to us; they *are* us.

DISTANT REFLECTIONS IN A NEARBY UNIVERSE

We couldn't keep from wondering if we might somehow be seeing
our own origins.

—Robert E. Williams, former director of the Space Telescope
Science Institute and Hubble Deep Field team leader, 1995,
upon viewing the Hubble Deep Field for the first time

In late November, as darkness falls, a huge four-starred asterism
called the Great Square of Pegasus floats high overhead in the
Northern Hemisphere; in the Southern Hemisphere you will find
it low in the northern sky. If you let your eyes drift north-northeast
of the star occupying the square's northeastern corner, and if your
skies are dark enough, you will see an oval misty glow that you
could cover with the tip of your outstretched thumb. This pale
presence, unimpressive though it may appear to the eye, is our
galaxy's nearest comparably sized neighbor, the Andromeda galaxy.
With a distance of a little more than 2 million light-years, it qualifies as one of the most distant objects visible to the unaided eye.

Images that capture the full width and breadth of this galaxy
show a slightly warped disk rimmed with dust surrounding a

broad, pearly center. The impression is that of a smooth whirlpool of light spinning as leisurely as a carnival carousel. Apparently, though, the galaxy wasn't always this placid. In late 2002 and early 2003 astronomers used the Hubble Space Telescope to sift through the stars in the Andromeda galaxy's halo. The halo is an all-encompassing shell of Sun-like stars and globular star clusters dating back to the early universe—at least that's what astronomers had always assumed based on the ages of halo stars in our own galaxy, which fall between 10 billion and 13 billion years. With the Andromeda galaxy virtually next door, and hence in the same evolutionary neighborhood, as it were, they expected to see similarly aged stars in its halo. Much to their surprise, however, they discovered that fully one-third of the halo stars they examined are far younger, between 6 billion and 8 billion years old. Something had come along in the galaxy's past and reactivated its star-making machinery.

That "something," astronomers think, was a merger between the Andromeda galaxy and one or more galaxies billions of years ago. The collision, or collisions, may have deposited younger stars into the galaxy's halo, dislocated younger stars from its disk, or created new stars as the two systems merged into one. In any case, it appears that a galaxy's halo may be something more consequential and dynamic than a quiescent stellar sanctum where old stars grow ever older and not much else happens. Perhaps halos are a kind of palimpsest of galactic evolutionary history written and erased many times over.

Galaxy mergers, though rare today, were more common in the early universe when galaxies were closer together and gravity could have its way with everything. Indeed, the Hubble Space Telescope image of the Andromeda halo reveals myriad galaxies lying billions of light-years beyond. Many have distorted or amorphous shapes, semaphores of violent collisions. The blending of galaxies in this Mixmaster phase of the cosmos is what has largely endowed our universe with its present form and content. The ages and chemical contents of the stars and of our Sun, the formation of planets, and the emergence of life all precipitated from the interactions and evolution of the first stars and galaxies.

Though the thrust of this book has been what astronomers have learned about the first stars and galaxies in the early universe, we cannot overlook the fact that we live in a galaxy, too, and are surrounded by galaxies that are only millions, not billions, of light-years away. Just as the Sun makes an excellent laboratory for studying the nature and evolution of more distant stars, so, too, can the properties of our galaxy and its neighbors be correlated to galaxies in the deep end of the universe.

One approach is to study the ages of the Milky Way's oldest stars, called white dwarfs, and use them as stellar chronometers to obtain the age of the universe. Such stars are ideal for this purpose. White dwarfs are the cindery remains of stars that originally had masses no more than eight times that of the Sun. A star like our Sun can live approximately 10 billion years, while those with lower masses can live even longer. Thus many of the white dwarfs now smoldering to stellar ash in our galaxy could very well be among the first stars to have formed in a much younger universe.

Recently, UCLA's Bradley Hansen and his colleagues set out to determine exactly how old our galaxy's crop of white dwarfs are. Using the Hubble Space Telescope, they focused on the globular cluster M4 in the constellation Scorpius. Globular clusters are popular objects with amateur astronomers because many are bright enough to be visually spectacular in small-to-medium-sized telescopes. But Hansen wasn't interested in M4 for aesthetic reasons. Globular clusters are the stellar repositories of the oldest stars in the galaxy, including white dwarfs. But because globular clusters are notoriously distant—most are over 12,000 light-years away—their contingents of white dwarfs are too faint to observe. M4, however, is the closest globular cluster to Earth—7,000 light-years—and thus even feeble white dwarfs are fair game for the HST.

Hansen's team selected a field along the less-dense outskirts of the cluster to avoid overcrowding and made exposures totaling over 177 hours, making it the deepest exposure of any star cluster ever made. The resulting image revealed a dazzling spray of diamond-bright stars. Scattered among these brilliant crystals were white dwarfs, mere specks and some barely visible.

The age-dating system Hansen and his team used is analogous to estimating how old a fire is by measuring the temperature of its coolest embers. In the case of M4, white dwarfs comprise the embers, the coolest being the oldest. In order to do this with any degree of confidence, though, they had to run through a lot of white dwarfs.

"What we rely on is the fact that the rate of white dwarf cooling is not constant," says Hansen. "The cooling rate depends on the temperature of the white dwarf, its mass, what it has in its atmosphere, and so on. Now, for a single white dwarf there are several parameters, and so you can tweak a parameter here and there and get a variety of ages that fit the observations. But once you have a couple of hundred white dwarfs, making your model fit all of them at once means that you start to constrain the model quite severely, and this is basically what we do. With enough white dwarfs and good enough observations, only a limited range of ages will fit the data."

The team found that some of the white dwarfs in M4 are truly hoary stars—between 12 billion and 13 billion years old. That means the stars populating at least some of the globular clusters in our galaxy formed when the universe was a billion years old or less. Their measurements fall nicely in line with the best age estimate of the universe to date—13.7 billion years—while also offering a completely independent way of pinning down that value.

Another intriguing aspect of their work is that it enables them to assess the ages of our galaxy's halo and disk. It has long been thought that globular clusters accompanied the formation of the galactic halo and bulge. After several billion years, infalling gas from these structures settled into a hot, thick protodisk that resulted in a modicum of star formation. Only after the disk cooled still further, which took another few billion years, did a true thin disk form, the one we have today.

Applying the age derived for M4, Hansen and his team found that our galactic halo was in place over 11 billion years ago, around a redshift of 3, and the galactic disk assumed its present thin form some 8 billion years ago, around a redshift of 1.5. "That doesn't mean it looked like it does today," says Hansen. "A common occurrence at early times is the merger of two or more already formed galaxies. Such events can cause quite a mess. Many believe this is what caused the thick disk in our galaxy, which is a flattened structure that rotates like a disk but is thicker and only about 10 percent as massive as the current thin disk. It is thought this material started forming a disk, but the Milky Way swallowed a sizable neighbor, maybe something the size of the Large Magellanic Cloud, which stirred it up to make the thick disk. The thin disk then formed from gas that had not yet formed into stars at that point."

Knowing the ages of the oldest stars provides a conclusive timeline for the formation of our galaxy, but plotting the motions of stars toward and away from Earth—a component called radial motion—can provide a dynamic history. The more stars you can track, the better picture you get of star "streams," which represent the remnant pathways of small satellite galaxies that interacted with the Milky Way billions of years ago. The number and trajectories of such streams indicate how the various Milky Way components initially assembled themselves, as well as provide insight into how other galaxies might have formed.

For example, recent evidence suggests that dwarf satellite galaxies were stripped of stars as they passed through the dark matter halo of their larger host. Paul Martini, of the Harvard-Smithsonian Center for Astrophysics, and Luis Ho, of the Carnegie Observatories, measured the motions of fourteen globular clusters associated with the giant elliptical galaxy NGC 5128 (also known as Centaurus A), in the southern constellation Centaurus. They found that the properties of these clusters—their luminosity, structure, and orbital motions with respect to the galaxy—overlapped quite well with the central properties of compact dwarf galaxies. Six of the globulars also exhibited tidal tails, material that is inevitably drawn out of a small galaxy when it passes near or through a larger galaxy. Computer simulations have suggested that this kind of galactic abrasion can transform the satellite from one type of dwarf galaxy into another, as well as create the kind of star streams observed in the halos of the Milky Way and Andromeda galaxies. Martini and Ho, however, suggest that some of the pithy remnants of these tidally stripped dwarf galaxies are, in fact, globular clusters.

"The whole globular cluster population of a giant elliptical like NGC 5128 has implications for galaxy formation because globular clusters provide insight into the merger and assembly history, particularly if new generations of globular clusters are formed by major interactions between galaxies," says Martini. "For the most massive globular clusters, such as those in our study, they are of particular interest because of the potential connection to dwarf galaxy formation. Tidally stripped dwarf galaxies could produce objects that are morphologically very similar to these massive globular clusters. Thus this population might also be valuable for understanding the dwarf galaxy population."

Martini and Ho conclude that massive globulars like those around NGC 5128 and less-massive globulars like those in the Milky Way are not wildly different objects. But they are still a long way from knowing whether there's a real relationship between the two or just an astrophysical coincidence. "More data are needed to determine if these objects are stripped dwarfs or simply the most massive star clusters," says Martini. "Such observations would include making deeper images to determine if they have relic envelopes from their dwarf galaxy past."

Although a tidally stripped dwarf may look like a globular, most dwarf galaxies are extremely faint and hence difficult to detect. In October 2004 astronomers with the Sloan Digital Sky Survey announced the discovery of a new dwarf galaxy in the Milky Way halo. The dwarf, dubbed Willman 1 after team leader Beth Willman of New York University, was detected as a slight enhancement in the number of stars occupying a small region of sky. All told, the dwarf is some two hundred times less luminous than any known galaxy, making it nearly transparent. But in a statement that resonates with Martini and Ho's work, Michael Blanton, also with New York University, says that Willman 1 might be an unknown type of globular cluster. "Its properties are rather unusual for a globular cluster," he says. "If it is a globular cluster, it is probably being torn to shreds by the gravitational tides of the Milky Way."

More than likely, Willman 1 is one of the many "missing" dwarf galaxies the cold dark matter model predicts the Milky Way should possess. Currently only eleven dwarf galaxies are known to be orbiting our galaxy. There should be hundreds. "If this new object is in fact a dwarf galaxy," says Willman, "it may be the tip of the iceberg of a yet unseen population of ultra-faint dwarf galaxies."

As mentioned earlier, when smaller galaxies merge with a larger one, they often create star streams as a result of their gravitational encounter. Different galaxies produce different groups of star streams. Brad Gibson of the Swinburne University of Technology is leading an ongoing project called the Radial Velocity Experiment, or RAVE, to look for them. He and his team use the 1.2-meter UK Schmidt Telescope in Australia to make radial velocity measurements of 100,000 stars, from which they may be able to identify dozens, perhaps hundreds, of streaming star groupings. Gibson refers to his group as "galactic archaeologists."

"Measurements obtained from the telescope will give new insight into how our Milky Way galaxy formed," Gibson says. "Our team's galactic archaeologists' sifting through the wealth of historical RAVE data will, for the first time, undertake a star-by-star reconstruction of our Milky Way."

Phase II of the project, which will run from 2006 to 2010, will measure 50 million stars. Regions will be sampled in the thin and thick disks, bulge and halo, looking for "schools" of streaming stars.

Astronomers also draw upon a technique called comparative morphology to contrast the size and shape of the Milky Way to galaxies at successively greater distances to gain a greater understanding of how galaxies change over time. That's the approach UCLA astronomers Tiffany Glassman and James Larkin took in a recent study. They observed ten galaxies about 5 billion light-years away, which corresponds to a period when the universe was about two-thirds of its present age. This time span represents a significant chunk of a galaxy's past, far enough back in time to help astronomers discriminate between differing theories of galaxy formation. They found that the disks of distant spiral galaxies are 2.5 times brighter at their center but 20 percent smaller than local disks and the Milky Way. These results support other observations that indicate that large galaxies were assembled from smaller ones more than 10 billion years ago and are slowly fading as the rate of star formation declines.

Looking within the galactic neighborhood, astronomers are seeing additional corroborating evidence for this evolutionary turning point. Astronomers Ben Panter and Alan Heavens, of Edinburgh University's Institute for Astronomy, and Raul Jimenez, of the University of Pennsylvania, analyzed the overall color of forty thousand nearby galaxies from the Sloan Digital Sky Survey (SDSS). Galaxies shine with the combined light of all the stars within them, producing a continuum of color that can be used to determine their rate of star formation. Starlight coming from massive young stars is blue. After only 10 million to 100 million years, however, such stars die out in a blaze of supernova glory. After their incineration they no longer contribute to the light of the galaxy. Instead, older and longer-lived redder stars do. Hence the redder a galaxy's stellar population appears, the older it is likely to be. Says Jimenez, "Our method takes into account all the stars that are present in the observed galaxies today and allows us to create the most complete history of star formation yet assembled."

In their overall analysis, the team found that of the forty thousand galaxies they sampled, most looked predominantly red rather than blue. This finding, they say, is a sign that the ascendancy of star formation is lessening not only for these galaxies but also for the universe.

"Our analysis confirms that the age of star formation is drawing to a close," says Heavens. "The number of new stars being formed . . . has been in decline for around six billion years—roughly since the time our own Sun came into being."

In follow-up research a year later the team examined nearly 100,000 SDSS galaxies. Their findings showed that the mass of big galaxies, such as ellipticals, has not changed since a redshift of 0.35, which corresponds to about 4 billion years ago, thus supporting the declining star-formation rate found the previous year. They saw something else, too: a clear correlation between early star-formation activity and present-day stellar mass. Some 7 billion years ago, by a redshift of 1, massive galaxies had rapidly converted 80 percent of their gas reservoirs into stars. Less-massive galaxies such as spirals, however, were more miserly with their raw materials. By a redshift of 1 they had converted only about 20 percent of their gas into stars. They would take another few billion years to transform the rest.

The researchers' conclusion was that dark matter halo formation, which is hierarchical in nature—that is, it helps build up smaller galaxies into bigger ones—is actually antihierarchical when it comes to forming stars. Massive galaxies apparently transform more gas into stars at higher redshifts, while less-massive galaxies transform more gas into stars at lower redshifts. This, says Jimenez, strongly suggests that massive galaxies were among the first to form stars at high redshifts and that dark matter halos amassed at the same time as the galaxies' stellar population.

"Our findings imply that elliptical galaxies dominated the early history of star-formation history," he says. "We *do* need dark matter halos. In fact, we find that there are sufficient dark matter halos at high redshifts to harbor these massive galaxies. This is in very good agreement with the current cosmological paradigm."

And once again it also indicates that star formation will never again be as robust as it was billions of years ago, around the time the Sun formed. "The universe is on the decline," says Jimenez. "Most of the gas has already been locked up into stars, and the future star-formation rate will be at the same level as is found today, not larger."

What is evident from these observations is that the properties of today's galaxies translate fairly well to those existing at earlier epochs. But, says Andrew Bunker of the Institute of Astronomy at Cambridge University, comparative morphology is not without its uncertainties. One of them is the way a galaxy shows a different side to itself when it lies at very great distances. The reason: the spectrum that it would exhibit if it were not subject to cosmic expansion—in other words, its "rest-frame" spectrum—has been redshifted from ultraviolet to red wavelengths.

"If you are observing a distant galaxy sample at the same wavelength as a nearby sample," says Bunker, "then you are actually seeing further into the rest-frame ultraviolet in the distant galaxies because of the redshift caused by expansion. We know that the same galaxies viewed at different wavelengths show different structure. In the rest-frame ultraviolet you are dominated by recent star formation, which tends to be clumpy, star-forming knots. But in the near infrared you are sensitive to older stars, and the overall stellar distribution is smoother."

Another pitfall, Bunker says, is that the different components of galaxies conspire to make them grow dimmer in different ways out to cosmological distances. "A galaxy at a redshift of 1 will have a surface brightness sixteen times fainter than one at a redshift of zero," he says. "But this applies to resolved regions of light. If you have an unresolved source, perhaps a quasar or compact star-forming region, then the dimming is less extreme. So this has a tendency to boost the sensitivity to these regions in distant galaxies, making them look different."

Other unknowns involve how these galaxies evolve in color and size over time and how their mass changes as a function of redshift. Any one of these factors could potentially introduce systematic errors that, in turn, would skew astronomers' conclusions about the nature of high-redshift galaxies.

A new technique called cloning may resolve these issues. The idea is to take images of nearby galaxies and artificially redshift them out to distances equivalent to galaxies in the Hubble Ultra Deep Field. The procedure, developed by Rychard Bouwens of the University of California at Santa Cruz, can be used to determine how galaxies are distributed in surface brightness, color, size, form, and abundance as a function of redshift.

"Cloning is perhaps the most elegant and powerful way to test different hypotheses regarding the evolution of galaxies from one redshift to another," says Bouwens. "Instead of representing a population of galaxies by a set of model profiles, colors, and so forth, you simply represent the population by all the members that make up a given sample. The trade-off . . . is that it is necessary to archive the properties of all the galaxies in different samples and have the software project them to different redshifts."

Bouwens begins by color-correcting each pixel of a galaxy's image to reflect how the galaxy would appear at the longer wavelengths found at a particular redshift. Next, he scales each pixel both in size and in brightness to reflect its increased redshift value, then smooths the overall image to compensate for its smaller size. Finally, he introduces instrument noise into the image frame, since that, too, is a big factor in how well galaxies are resolved at high redshifts.

The simulations provide a remarkable analog in color, brightness, and morphology to real high-redshift galaxies. But the technique also promises to circumvent a major impediment facing astronomers in analyzing high-redshift galaxies. As mentioned, the light seen in galaxies like those in the Hubble Ultra Deep Field has been "downshifted" by expansion from ultraviolet wavelengths to lower wavelengths; thus that is what predominates in the visual images—ultraviolet light. The knotty, irregular appearance of these galaxies is therefore likely due to internal spates of star formation. But the light of mature, established stars is "off the scale" because it, too, has been redshifted to longer wavebands that are beyond the reach of conventional instruments. Therefore it's not possible to create a complete picture of star formation in high-redshift galaxies.

Complicating this challenge is the fact that the ultraviolet aspects of normal nearby galaxies are not well understood anyway; thus it is difficult to draw complementary structural conclusions to those at high redshifts. The question then becomes, to what extent do the large number of knotty blue galaxies seen in the Hubble Deep and Ultra Deep Fields result from redshifted ultraviolet light rather than pure evolution? Galactic cloning may be able to answer this question.

The cloning application also provides a model-independent method for measuring the evolution of galaxies from high redshift to low. The technique has already been fruitful in confirming that galaxies in the Ultra Deep Field between redshifts of 2 and 6—a difference of about 2 billion years—are predominantly compact in size (less than 10,000 light-years across), while large low-surface brightness galaxies are rare.

Bouwens is currently working on an ambitious technique for backwardly evolving galaxies from low to high redshifts in an effort to enumerate how individual galaxies evolve throughout the history of the universe. His approach is holistic, factoring in how age, dust, and metallicity affect the colors of stars in galaxies, as well as what is known about stellar evolution, galactic mergers, and internal dynamical changes such as the formation and effects of the propeller-shaped bars in spiral galaxies. Galactic cloning and "backward evolution" may eventually enable astronomers to watch galaxies form and evolve over a wide range of redshifts.

That's a fascinating prospect, to be sure. But it is all the more remarkable given that it has been only a decade since the study of how galaxies evolve was severely challenged by the weird objects cluttering the Hubble Deep Field. Now for the first time theory and observation are allowing astronomers to trace the evolution of galaxies almost step by step from the modern to the primordial universe. We can run the galactic evolutionary movie backward to see how a galaxy like our own can morph from its present star-encrusted form into one with a thick, hot disk, fewer stars, and a rudimentary halo. We can watch as massive elliptical galaxies, which have glowed for aeons with the soft light of Chinese lanterns, flare up and become ablaze with stars, while others fragment into a half-dozen or more smaller knots of light before dispersing like fireflies in the growing darkness. To some degree, we can follow the stages of galaxy formation right into the fog bank of the Dark Ages and beyond, even toward the big bang itself.

In January 2005 astronomers with the Sloan Digital Sky Survey and the 2-degree Field Galaxy Redshift Survey (2dFGRS) assembled in San Diego at a meeting of the American Astronomical Society to announce that both groups had detected ripples in the distribution of galaxies made by sound waves generated soon after the big bang. Moreover, the sound waves, which were generated by density disturbances present in the plasma after the big bang, can be traced to modern galaxies like the Milky Way.

The 2dFGRS team, composed of astronomers from the United Kingdom and Australia, identified the imprint of sound waves using a statistical device called a power spectrum. Don't let that term put you off. A power spectrum is simply a plot showing the power range of a signal over various frequencies. White noise, similar to what you get with an untuned television channel, is a signal whose power spectrum is flat across frequencies. The cosmic microwave background, too, can be considered a flat, white-noise spectrum—that is, until you begin looking at it on angular scales of a degree or less, where enhanced signals, or acoustic peaks, emerge.

In terms of the distribution of galaxies, the power spectrum plots the clustering of galaxies throughout various regions of physical space. At the largest scales in the universe, and in accordance with the cosmological principle, the distribution of galaxies looks white-noise flat, or homogeneous. But when the distances and positions of a lot of galaxies are measured over a great volume and to fine enough scales, the "signals" of discrete structures—galaxy clumps—appear. These can then be compared to the acoustic patterns plotted in the CMB. Over a ten-year period the 2dFGRS team measured the redshifts of thousands of galaxies per night—a total of 220,000. Analysis of the resulting power spectrum revealed that the pattern of acoustic peaks in the cosmic microwave background correlates to the galaxy clumps measured by the 2dFGRS team.

"We see the structures in the local galaxy distribution convincingly for the first time," says Richard Ellis, who acted as spokesperson for the 2dFGRS group. "They are on the expected scale."

The SDSS team took a different approach. They mapped more than 46,000 galaxies (about 10 percent of the northern sky) out to nearly 5 billion light-years. The astronomers then began looking for clusters of galaxies separated from one another by 500 million light-years, which is the predicted scale these acoustic ripples would have after propagating outward for a million years from an initial disturbance in the primordial cosmic "lagoon." Their working assumption was that more galaxies were likely to be separated by 500 million light-years than by 400 or 600 million light-years. This was exactly what they found.

Says lead investigator Daniel Eisenstein of the University of Arizona, "We regard this as smoking gun evidence that gravity has played the major role in growing, from the initial seeds in the microwave background, the galaxies and clusters of galaxies that we see around us."

As an added bonus, the results are in precise accord with the predictions of the standard cosmological model, in which all the mass in the universe is dominated by dark energy. Commenting on these results, Sir Martin Rees of Cambridge University says, "The concordant picture we have of the universe is hanging together very well. In a way, it might be more exciting if we found a glaring inconsistency, but it does seem that we *have* consistency, more than we expected, and the standard picture is firming up. . . . We're coming to understand how over the last 90 percent of cosmic history, structures started forming galaxies and those evolved to what we see today."

All we used to know of the universe was what could be seen with our limited vision. It must have been frustrating for Edwin Hubble to follow the landscape of galaxies deeper and deeper into the darkest reaches of the universe, until there was no more light to find, through no fault of man or science. "The history of astronomy is a history of receding horizons," he wrote in *The Realm of the Nebulae*. The horizon Hubble most wanted to breach was the farthest of those: "We know our immediate neighborhood rather intimately. With increasing distance, our knowledge fades, and fades rapidly."

Now breakthrough telescopes, sophisticated photon detectors, and computer models have enabled astronomers to bridge that distance and bring to light the remarkable vistas that Hubble could only imagine. What would he think of it all? No doubt he would marvel at the array of malformed galaxies and their evolutionary implications, but he would also point out that there are yet more shadows and receding horizons to pursue.

MORE OF EVERYTHING

With all due respect to astronomers: We don't NEED to find any more stuff in the universe. We already have more stuff than we could ever use, right here in our garages.

—Dave Barry

"The difficulty of writing a book like yours," Rodger Thompson said to me with his knowing grin, "is that there is no end to a book like yours." To which I readily admit. And yet, if nothing else, it's worth providing some record, however mercurial, of the impressive gains made over the last decade in the study of this empyrean armature we call the universe. For nearly a century the field of observational cosmology has slowly but surely advanced. That's not to say that time hasn't been spent drifting in a Sargasso Sea of stalemate or storm-cast controversy, but when a new bearing was laid aloft, it was always forward, driven by the winds of consensus. Today there are only a few isolated pockets of dissension, but no one has any delusions about the challenges cosmologists and astrophysicists face.

WMAP researcher Gary Hinshaw is not alone when he says, "Determining the nature of the dark energy is hands down the biggest challenge facing both observational cosmology and fundamental physics as well. We are faced with the prospect that 70 percent of the energy density in the universe is locked up in a form that is *very* difficult to observe directly. It required a massive effort just to determine how much there is, let alone what its detailed properties are. But it has such profound implications for physics beyond the standard model, and it's one of our few clues about such physics, that it's worth the effort."

In the early universe, dark energy was less a player than dark matter—yet another challenge—but it apparently began influencing large-scale structure around a redshift of 1 or 2, which is why mapping out its crenulated edges is so important. The signature of dark energy may have already been detected in the brightness diminution of Type Ia supernovae at redshift distances of 0.5. Being 10 to 20 percent *fainter* than astronomers expected in a universe that was slowing down due to gravitational drag, something—read: dark energy—must be causing it to accelerate to make them appear fainter than predicted. Additional X-ray and microwave observations of galaxy clusters independently support this finding.

But to understand the nature of dark energy, particularly if it varies in pressure and density over time and distance, many more measurements of Type Ia supernovae need to be made. The hope was that the Hubble Space Telescope would be instrumental in providing these measurements until the end of the decade, but given its almost-certain untimely demise, the course heading in this direction is unclear. As of this writing, calls for observing proposals are still being made at the Space Telescope Science Institute, and astronomers are responding, though somewhat fatalistically. One of them is Adam Riess, whose work on measuring the nature of dark energy via Type Ia supernovae would be severely affected.

"I'm thinking, 'How can we squeeze as much out of the next year or two as we can?' " he says. "I have to face the reality that we're probably not going to be able to go for very long. It's hard to predict when things will fail on the telescope. We have all these probability curves that tell us our best guess as to when the telescope will fail, but it's like predicting how many miles you have left on your car. You think probably at about 100,000 miles, it starts getting dicey, but it could keep running to 150,000 miles."

Right now the Hubble Space Telescope is one of the best tools astronomers have for determining the current strength of dark energy and how it might have changed over time. "To do that," says Riess, "we measure the most distant supernovae, above redshift 1. These are part of the set of clues that we can obtain about the dark energy—what it is, whether it's been changing, or whether it's a permanent feature of the universe. We don't know how long the telescope will last, but if it fails . . . it will leave a definite hole in our tool kit for making these kinds of measurements."

After Hubble, says Riess, astronomers will have to wait until a new supernova telescope, called the Joint Dark Energy Mission, or JDEM, is placed in orbit. Formerly known as the Supernova Acceleration Probe, JDEM is a cooperative venture by the Department of Energy and NASA that will study thousands of supernovae out to a redshift of 1.7 with unprecedented precision, using a wide-field 2-meter telescope operating from optical to near-infrared wavelengths. These observations should afford astronomers the means to distinguish between different models that describe how the density of the universe evolves over time, putting scientists one step closer to understanding the nature of dark energy. The mission is not expected to be launched, however, until 2015.

In the near term a project called the Supernova Legacy Survey is expected to deliver more than a thousand Type Ia detections, which should go a long way toward narrowing dark energy's "equation of state." That study will also provide comprehensive observations of how reddened Type Ia supernovae appear at various distances due to intergalactic dust, something that currently is not understood with any degree of accuracy. On the downside, the legacy survey will be able to detect only supernovae with redshifts less than 1. "This probes only the recent history of dark energy and tells us about its current strength," says Riess, "but not so much about how it's been changing over time. I think we're going to need both [high- and low-redshift measurements] if we're going to understand what dark energy is."

If dark energy is responsible for increasingly pushing galaxies apart in the "Middle Ages" of the universe, then dark matter is what began mustering matter in the Dark Ages following the big bang. But dark matter notwithstanding, *all* structure in the early universe fell out of the density irregularities present in the cosmic microwave background as slight temperature variations—*very* slight, one part in a million. These are the veins within which the rich ore of the first galaxy clusters was deposited. The COBE and WMAP probes were successful in mapping these riffled regions, but cosmologists know that the CMB holds a lot more information yet to be charted and constitutes yet another imposing challenge for researchers.

For the immediate future, cosmologists are investing great expectations in a European Space Agency satellite called Planck that is expected to be launched in 2007. Named in honor of German physicist Max Planck, the author of quantum theory, the probe's objective is to provide the most precise measurements possible of the CMB's temperature variations. With three times the resolution of WMAP, Planck should have no trouble meeting that goal.

But another probe, still in the planning stages (and thus highly vulnerable to imminent cancellation should the funding evaporate), would be able not only to nail down questions about the emergence of structure but also reveal a great deal about the physics of inflation. Called the Beyond Einstein Inflation Probe, its mission will be to look for traces of gravitational waves imprinted on the relict radiation of the cosmic microwave background. The idea here is that the sudden acceleration of expansion that occurred during inflation should have created gravity waves that would have rippled throughout the primordial plasma of the just-born universe like the waves caused by a brick tossed into a quiet pond. These gravity ripples, in turn, would have polarized the light of the cosmic microwave background. Hence primordial gravity waves should be detectable in the polarization patterns in the background radiation.

Polarization is an either/or condition, as in politics when opposing factions are said to be polarized, in magnetism where the opposite ends of a magnet exert forces in opposite directions, and in geometry where either end of the axis of a sphere is considered a pole. In optics light is said to be polarized when light waves are made to propagate in one direction, either by reflection or by scattering. That is why polarizing sunglasses are so effective at cutting down on glare. Usually ambient light waves vibrate in all planes perpendicular to their direction of propagation. In other words, they come at you from many different sources and along random planes. But light becomes polarized whenever it reflects from a car windshield or off the surface of a lake or pond. Polarizing sunglasses are like optical gatekeepers that block the passage of vibrating light waves in one direction but permit vibrating waves in the perpendicular plane to pass.

Polarization of the CMB has long been predicted, and in 2002 it was finally detected by the Degree Angular Scale Interferometer

(DASI), situated at the National Science Foundation's Amundsen-Scott South Pole Station. Additional confirmation was supplied in 2004 by two years' worth of observations made with the Cosmic Background Imager, located high in the Chilean Andes, and clinched that same year by the first release of the WMAP data. The results of these independent observational approaches tell the same story. The cosmic microwave background became polarized when the microwave photons scattered off of the first atoms 400,000 years after the big bang, producing more photons coming from one direction than another. Before this time no polarization pattern can be seen. The universe was too hot then to allow electrons to combine with atomic nuclei; hence photons were constantly ricocheting off electrons without exhibiting a preferred polarization pattern. Only after the universe cooled and atoms formed did rampant photon scattering dampen out and a polarization pattern emerge.

The Inflation Probe—should it ever fly—will take CMB polarization one step further by more closely examining whether inflation took place at temperatures comparable to the energy scales expected in Grand Unified Theories, in which the four fundamental forces of nature—the strong, weak, electromagnetic, and gravitational forces—are fused into one. If such inflation did take place, says WMAP's Hinshaw, then the remnant gravity waves produced during inflation would have left a measurable imprint in the polarization of the CMB. "This would give us a direct probe of physics on grand unified scales—twelve orders of magnitude higher than the energies to be achieved by the Large Hadron Collider being built at CERN," he says.[*] "The ties between cosmology and fundamental physics are rich indeed."

And, one might add, inevitable. Working as one, the two fields

---

[*]CERN is the European Centre for Nuclear Research.

have already produced a bounty of new insights. If the alliance continues, as it must, it will profoundly affect the way research in astrophysics and particle physics is conducted and, in so doing, render a more comprehensive picture of the universe. In this seemingly antithetical realm, the gradations between quarks and stars, atoms and galaxies, will become more indistinct as the role each plays in endowing the universe with its observed properties becomes better understood.

For all that is already known about the universe, the future of cosmology apparently holds more—more cross-pollination of fields, more telescopes, more wavelength coverage, more sophisticated instrumentation, more dynamic theoretical models. Even though observers using existing telescopes are detecting galaxies at redshifts of 7 and 8, observers continue looking to the day when a new generation of space telescope or supercolossal ground-based telescope is brought to bear on both the near and far universe. It's not that observational astronomers don't appreciate what they have in the way of technology. They want more because they know so much more is out there.

Astrophysical theorists want the same thing, and Michael Shull of the University of Colorado confirms this point. "I ask my wife a similar question: 'What do women want?' Her answer is, 'They want more.' Same for me. I want answers to my existing questions, but more important, I want challenging new observations, paradoxes, puzzles to explain. Those puzzles usually come from better observations, meaning opening up new wavelength bands, higher-quality data, and sometimes just a new way of reanalyzing existing data. A clever observer often does just that."

Shull would like to see a "true" successor to the Hubble Space Telescope, perhaps a 6-to-10-meter ultraviolet/optical telescope capable of imaging and taking ultraviolet spectra of galaxies and gas between the galaxies to faint magnitudes. "Optical and ul-traviolet astronomy are due for a major increase in observing power," he says. "Hubble did so much to study the universe with a relatively small 2.4-meter space telescope. Just imagine what a 10-meter could do!"

Galaxy morphologist Christopher Conselice lands squarely on the need for innovations in observing technology. At the top of his wish list are more infrared detectors. These would enable galaxy morphologists to coax out fainter structures in the earliest galaxies. "We seriously need large imagers on 8-to-10-meter telescopes and eventually on 20-to-30-meter-class telescopes," he says. "Real advances will be made when multi-object infrared spectrographs are built to carry out spectroscopic surveys in the near infrared." The James Webb Space Telescope, he adds, will be a critical step in the right direction: "It will greatly advance the field because it will allow us to probe many times deeper at infrared wavelengths than we can do now."

Conselice also resonates with Shull's interest in exploring novel ways of thinking about and applying existing data, like the Hubble Deep and Ultra Deep Fields. Such new approaches could potentially reap new and interesting science, he says. "In terms of morphology work, we still need to think carefully about the best way to characterize galaxy structure and to devise a system whereby galaxies are related to each other in a physical way. This is in many ways more important than the technical advances. We can now study nearby galaxies at high resolution at all wavelengths in detail, and high-redshift galaxies in the optical and near infrared. The data is there; we just need to think more carefully about how to use it."

Since the days of Galileo, astronomers have always wanted more of everything. Once a telescope of any size was built—a 40-inch, a 60-inch—the dream was always of building an even bigger one—a 100-inch, a 200-inch. When photographic plates came along, astronomers pushed to make more red-sensitive emulsions. The advent of electronic light detectors revolutionized observational astronomy, but they always could be more sensitive, take in larger fields of view, and be less susceptible to background noise. Fortunately, over time technology succeeded in realizing those dreams, and in turn, our knowledge of the universe was advanced by observational developments arising from those telescopes.

The Hubble Space Telescope put a different spin on what astronomers wanted. Of necessity, since it was going to be launched on the space shuttle, it had to be of modest size and not overly equipped with instruments. So while it wasn't a bigger telescope, it could see more clearly than any on Earth. What astronomers got was sharper vision or, as they call it, better resolution. Resolution, or the ability of a telescope to detect detail in celestial objects, is determined largely by its light-gathering surface area. But the potential resolution of a large-aperture ground-based telescope is limited by Earth's dense and quavering atmosphere. This is why the best observatories are located on mountaintops and extinct volcanoes. But put even a modest-sized telescope into space, and suddenly features barely glimpsed in the largest Earth-bound telescopes are revealed in stunning detail. This was the dream realized by the Hubble telescope.

Should Hubble not be serviced again in the near term, as seems likely, others will have to take up where it leaves off. One of those already contributing important observations is the Spitzer Space Telescope. This infrared-sensitive telescope is conducting deep surveys, including one centered within the northern Hubble Deep Field, probing galaxies out to distances of about 12 billion light-years, or about redshift 4.5. Another survey will cover about 70 square degrees of the sky out to distances of 10 billion light-years, or redshifts of 2. It is expected that this survey will net more than 2 million galaxies, some 30,000 per square degree. Besides its superior resolution at near-infrared wavelengths, another of Spitzer's great strengths is as a support platform for ground-based and space-borne telescopes. The telescope has a life span of about five years, however, and by decade's end a new generation of telescopes will have to take up the infrared slack.

Fortunately, two missions will do just that. The European Space Agency plans to launch the Herschel Space Observatory in 2007, and NASA is scheduled to launch the James Webb Space Telescope six years later, in 2013. If the Hubble Space Telescope can survive until after 2007, and barring protracted delays and disasters, the overlap will be almost perfect.

The Herschel Space Observatory will be the first space observatory to cover the full far-infrared and submillimeter wavebands, a region that so far remains unexplored. And with a mirror of 3.5 meters, it will be the largest of the second generation of space telescopes—that is, until the James Webb Space Telescope is deployed. By then, however, the Herschel mission will likely have come to an end, as it has a life expectancy of only three years. The baton will then be handed to JWST. With its 6.5-meter mirror and infrared-sensitive optics, this telescope will probe to unprecedented depths—predictions are of redshifts of 20 or 25, or some 180 million years after the big bang—well within the Dark Ages of the universe, where very early galaxies and stars lurk.

What will these primordial objects look like? More than likely they will not be as impressive as the confettilike galaxies strewn across the deep fields. "Initially, they'll see little red blobs of light," predicts stellar theorist Jason Tumlinson. "Of course, JWST will have extremely good spatial resolution so it will be able to see subtle differences in these blobs, but as you go further back, the galaxies, if you want to call them that, will get smaller, and eventually they may fade from view."

The heralds of these and the fainter "blobs" may be hypernovae—megasupernovae—the endgame of extremely massive stars. The earlier hypernovae may be only two or three times brighter than the supernovae seen in the universe today, but on the other hand, they will be twenty times more explosive. The reason they aren't twenty times brighter is because their progenitors are smaller. Says Tumlinson, "They're smaller objects and their radiating areas are smaller. Nevertheless, if they are two or three times brighter, you can still see them. We estimate you can easily see them at redshift 7 or 8 with JWST."

Red blobs aside, astronomers are largely optimistic as to how much detail JWST will reveal in early galaxies. Estimates are that the telescope's imaging sensitivity will be superior to that of a 30-meter ground-based telescope. That would be sharp enough to reveal hotbeds of star formation within nascent galaxies, as well as other morphological details such as tidal tails and mergers. Great strides should be made in areas such as the formation and evolution of galaxies and the physics of early star formation. Imaging and spectroscopic surveys should also pin down the epoch of reionization. Finally, by probing the gravitational fields of galaxies and galaxy clusters via weak and strong gravitational lensing, it should be possible to determine the distribution of dark matter on scales of individual galaxies up to galaxy clusters and beyond.

About a year before JWST goes into service, a radio telescope array currently undergoing construction in the Southern Hemisphere will be ramping up observations. If all goes as planned, the Atacama Large Millimeter Array, or ALMA, will consist of sixty-four 12-meter antennas located on the broad expanse of the Chajnantor Plain of the Chilean Andes, 5,000 meters above sea level.

(Site buildings will circulate oxygen-enriched air, and oxygen bottles will be available to astronomers and staff venturing outside.) The array, states one ALMA brochure, "will be a complete imaging, spectroscopic instrument" for millimeter and submillimeter observations. The antennas will be able to reconfigure from baselines ranging from a tight 150 meters across to 10 kilometers or more, providing a "zoom lens" capability for distant objects and a resolution five times better than the Hubble Space Telescope's Advanced Camera for Surveys.

The ALMA array will also have a redshift-frequency advantage when it comes to scouring the early universe. Recall that in an expanding universe, the light of galaxies at cosmological distances shifts from the ultraviolet through the visible and into the near-infrared regions of the spectrum. At higher redshifts, however, their emissions stretch through the infrared and into radio wavebands, beyond the reach of telescopes like the JWST. Because much of the dominant emission from the warm dust of nascent galaxies in the early universe is redshifted into ALMA's frequency bands, the array may be able to detect more distant galaxies, perhaps even true protogalaxies—those with little or no evolution that are still embedded in the Dark Ages beyond redshifts of 6—and perhaps out to a redshift of 20, when the universe was less than 200 million years old. This is a problematic era for galaxy formation, because the angular size of the density fluctuations in the cosmic background radiation indicates that galaxies are unlikely to have formed before redshifts of 20. It would be surprising indeed if galaxies with significant evolution are found approaching this redshift 20 boundary, because that would mean that protogalaxies would have formed much earlier than that.

Nearly a decade beyond both JWST and ALMA is SKA, or the Square Kilometer Array, which bills itself as "an international radio telescope for the twenty-first century." It truly is a global consortium comprised of fifteen countries and more than thirty participating institutions. As its name suggests, SKA's collecting area will be enormous—1 square kilometer, or 1 million square meters—some thirty times larger than the largest radio telescope operating today. The exact design and location of the SKA have yet to be determined, but the strawman plan proposed by U.S. astronomers is to use a whopping four thousand 12-meter antennas. The proposed locations are New Mexico, Australia, and South Africa.

Like ALMA, this radio giant will be able to map the structures of galaxies—given there are structures to map—with redshifts as high as 20, as well as follow the evolution of galaxy halos at redshifts less than 5. Covering frequencies of 0.15 to 20 gigahertz, SKA will be able to trace out the large-scale structure of "webs" of primordial gas, the sticky threads upon which condensed the first dewdrops of incipient galaxies. It will also detect smoldering star-forming regions in primordial galaxies and, it is hoped, glimpse the earliest pregalactic objects that formed in the universe during the Dark Ages.

SKA's highest frequencies will measure redshifted molecular lines in the interstellar medium of galaxies so distant they cannot be observed at optical wavelengths. Further, it will be able to gauge the rotational velocities of those galaxies, a key component in measuring their dark matter content.

If only half of these goals are met, the Square Kilometer Array will be a bracing success. As we have seen, anytime a new level of observing technology is attained, new discoveries are sure to follow. But with SKA there is an even headier possibility to consider: the potential for the discovery of a fundamentally new phenomenon in the universe. This expectation is not without a historical basis. The annals of radio astronomy are replete with serendipitous discoveries that have radically altered the course of science. For example, in June 1962 the moon happened to pass in front of a radio source cataloged 3C 273 in the constellation Virgo. Careful timing of the disappearance and reappearance of the object enabled radio astronomers at the Parkes Observatory in Australia to precisely pinpoint the source's position. Their subsequent radio maps of 3C 273 showed that it was, in fact, a highly compact object that coincided exactly with a faint blue star. That "star" would later prove to be the first known quasar.

And in 1967 graduate student Jocelyn Bell was using Cambridge University's 1.8-hectare radio telescope array to study the way radiation from the Sun affected signals from radio sources when she discovered "a bit of scruff" on the printout of the antenna's sweep of the sky. Follow-up observations revealed that the fixed source emitted a series of rapid and precisely spaced pulses. It turned out to be the first pulsar.

In addition to serendipity, a radio telescope's intended design has oftentimes been instrumental in making unintended discoveries. In a paper titled "Exploration of the Unknown," published in the December 2004 issue of *New Astronomy Reviews*, a team of radio astronomers led by P. N. Wilkinson of Jodrell Bank Observatory list a number of such instances. In 1933 Karl Jansky's "merry-go-round" array, which was meant to track the source of atmospheric static that was interfering with radio communications, was the first instrument to detect cosmic radiation emitted by the Milky Way. The Bell Labs Holmdel Horn, which serendipitously detected the cosmic microwave background radiation in 1963, was originally designed to measure continuum radiation of the Milky Way. And the majestic Arecibo dish in Puerto Rico, which has contributed to the confirmation of general relativity and the detection of binary neutron stars and gravitational radiation, as well as many other surprise discoveries, was intended primarily for the study of Earth's ionosphere.

So it is natural to wonder, what might an innovation such as the Square Kilometer Array turn up in wide samples of radiation gleaned from a universe only a few hundred million years old?

"What most radio astronomers study are very exotic things," says Ken Kellermann, a senior scientist with the National Radio Astronomy Observatory and one of six U.S. representatives to the International SKA Steering Committee. "Current radio telescopes can see quasars and powerful radio galaxies anywhere in the universe," he says. "The most powerful radio telescopes, like the Very Large Array near Socorro, New Mexico, can see galaxies with regions of active star formation and supernovae out to redshifts of 2 or 3, and normal galaxies, like our Milky Way, out to a redshift of 0.1 or so. But the main thrust of the SKA is its sensitivity, which requires a very large collecting area, not widely spaced antennas. With the SKA we will be able to study *normal* galaxies—if there are any—out to cosmologically interesting distances and to measure their redshifts. I don't know what it's going to discover, but it should resolve a number of current problems about galaxy and quasar formation."

Caltech astrophysicist George Djorgovski thinks that there are more fundamental new phenomena to discover in astronomy and cosmology, but that sooner or later, even with more powerful telescopes and greater wavelength coverage, we will run out of them.

"There is a finite volume of the observable universe, a finite range of the classes of things," he says. "Take the example of geography. You begin by not knowing anything about the size and scope of the planet Earth, but you find out: it is round, it is about this big, you count all the continents, all the mountains, rivers, and lakes. You map it out with satellites, lasers, radar, and sonar. You know about every geographical feature there is. You have a good understanding of how it got there in terms of the geology and whatnot, and how it will change, and that is all there is to it. There is no new fundamental geographical phenomenology to discover. The same thing will happen in astronomy and cosmology sooner or later. Our understanding of the universe will be complete in the same way as our understanding of the Earth's geography is now, even though our understanding of the universe may be now at a level, say, of geography at the time of Captain Cook. This is not to say that there will be no fundamental changes in our perception of the physical universe in the same sense as when quantum physics and relativity replaced the Newtonian view of the world. Such things can and do happen. But the phenomenology of the physical universe as currently perceived has to be finite in its scope."

Even if astronomers only encounter more rivers, trees, and mountains over the next celestial horizon, they will still have their hands full with the existing landscape of mysteries. Dark energy may be a priority now, but it isn't the only puzzle they have to piece together. Others involve the physics of the first objects to form in the universe. What, for example, are the signatures of the first stars? If the early universe turns out to be polluted with metals, the first true stars may reside in small clusters too embedded within the gloom of the Dark Ages to be seen, even with JWST. Age-metallicity questions abound with stars even closer to home, so this will be a challenge indeed.

How did the galaxies form? Even though astronomers can at last see galaxies in the process of forming, there are still significant gaps in their understanding. In order to draw hard-and-fast conclusions about their births, morphologists like Christopher Conselice and Simon Driver need to determine a complete "fossil record" of galaxies from those that exist today to those abiding at the highest possible redshifts. For galaxies at redshifts around 6, however, this is highly problematical. "We hardly know anything about them except that they are small and star-forming," cautions Conselice. "We cannot even say if they are merging, like redshift 3 galaxies. This is very much a work in progress."

Finally, three big questions rolled into one: When did reionization begin and end, and what sources were responsible? This is one of the more daunting challenges facing observational cosmologists, but it is one that holds the most promise for being answered in the near future. "We're very interested in understanding the physics of how the Dark Ages ended," says Caltech's Richard Ellis. "It is, in many ways, the last frontier in the origin of the galaxies."

Of course, as we have seen, the picture is complicated by the fact that, according to the WMAP data, the universe may have been reionized twice, first by Population III stars between redshifts 22 and 12 (150 million to 400 million years after the big bang, respectively), then again by low-metal Population II stars in large galaxies at redshift 6 (almost a billion years after the big bang). The numbers of burgeoning galaxies in the deep field observations would appear to support this scenario. If, says Ellis, you imagine looking at the density of star-forming galaxies per unit volume as a function of lookback time, "there seems to be a marked decline in their numbers. The Hubble Ultra Deep Field data is one of the most precise that has been so far measured. This marked decline seems to be important because the abundance is quite a lot lower than what would have been expected if this population we see were responsible for bringing the Dark Ages to a close. So we imagine that there must be earlier objects that together with this population are responsible for reionization."

Astronomers speculate that those earlier objects could have consisted of assemblies of luminous Population III stars, their hypernovae derivatives, some quasars, or a combination of all three. Reionization might also have been helped along by the accretion power of black holes. The disk of doomed material encircling a black hole is incredibly luminous due to the friction between layers of gas moving at different rotational velocities. On the other hand, how massive black holes could have formed so quickly and so early in the universe qualifies as yet another conundrum.

Sorting out these issues with any degree of confidence is an objective that younger generations of astronomers will have to pursue into their old age and likely hand off to a generation yet to be born. The search for knowledge, wrote Edwin Hubble over half a century ago, will continue, and "not until the empirical resources are exhausted, need we pass on to the dreamy realms of expectation." It doesn't take a dreamer, however, to conceive that, as far as knowledge of the universe is concerned, tomorrow and tomorrow and tomorrow will bring more and more and more: more comprehension, more mystery, more awe than we can imagine, and perhaps something new, not unlike the primordial objects captured in the Hubble Deep Field North and South, and in the Ultra Deep Field. Just one short decade ago nobody knew they existed, and certainly few astronomers, if any, were prescient enough to predict just how unusual their forms would appear.

Today astronomers are beginning to learn not only what these inscrutable objects reveal about the properties of the early universe but also how the universe will continue evolving long after humans have exited the cosmic stage. If nothing else, the search for the first stars and galaxies has shown that the future of the universe is indelibly inscribed in its past. The evolutionary history of our galaxy, ourselves, and other life-forms, should they exist, is writ there as well. The subtle script Edwin Hubble strove to read in the shadows of dwindling galaxies is gradually coming to light.

BIBLIOGRAPHY

Aside from interviews and e-mail exchanges with astronomers, a great deal of this book is based on the references below. The books listed provide mainly historical and fundamental background material, while the papers and articles reflect largely recent research, though a few, such as the letter by Fritz Zwicky, are of historical interest. With respect to the "astro-ph/" and "in press" references, since the production process of a book like this precedes its publishing date by roughly a year, readers may assume that most of these papers have now been published in science journals, books, or proceedings. Bear in mind that some of the final versions of those papers may differ slightly from the ones cited here.

BOOKS

Arp, Halton. *Atlas of Peculiar Galaxies*. California Institute of Technology, 1966. Available online at http://nedwww.ipac.caltech.edu/level5/Arp/Arp_contents.html.

Berendzen, Richard, Richard Hart, and Daniel Seeley. *Man Discovers the Galaxies*. Columbia University Press, 1984.

Danielson, Richard Dennis, ed. *The Book of the Cosmos: Imaging the Universe from Heraclitus to Hawking*. Helix Books, Perseus Publishing, 2000.

Eddington, Sir Arthur. *The Expanding Universe*. Cambridge Science Classics, 1987.

Harrison, Edward. *Cosmology: The Science of the Universe*, 2nd ed. Cambridge University Press, 2000.

Harwit, M. *Cosmic Discovery: The Search, Scope, and Heritage of Astronomy.* MIT Press, 1984.

Hubble, Edwin. *The Realm of the Nebulae.* Yale University Press, 1936, 1982.

Kirshner, Robert P. *The Extravagant Universe.* Princeton University Press, 2002.

Lang, Kenneth R., and Owen Gingerich. *A Source Book in Astronomy and Astrophysics, 1900–1975.* Harvard University Press, 1979.

Lemonick, Michael D. *Echo of the Big Bang.* Princeton University Press, 2003.

Livio, Mario. *The Accelerating Universe.* John Wiley & Sons, 2000.

Maran, Stephen P. *The Astronomy and Astrophysics Encyclopedia.* Van Nostrand Reinhold, 1992.

Münch, G., A. Mampaso, and F. Sánchez, eds. *The Universe at Large: Key Issues in Astronomy and Cosmology.* Cambridge University Press, 1997.

North, John. *The Norton History of Astronomy and Cosmology.* W. W. Norton & Co., 1995.

Peebles, P.J.E. *Principles of Physical Cosmology.* Princeton University Press, 1993.

Rees, Martin. *Just Six Numbers.* Basic Books, 1999.

Sandage, Allan. *The Hubble Atlas of Galaxies.* Carnegie Institution of Washington, 1961.

Shapley, Harlow, ed. *Source Book in Astronomy, 1900–1950.* Harvard University Press, 1960.

Silk, Joseph. *The Big Bang,* 3rd ed. W. H. Freeman & Co., 2001.

Singh, Simon. *Big Bang: The Origin of the Universe.* Fourth Estate, 2004.

Sparke, Linda S., and John S. Gallagher. *Galaxies in the Universe.* Cambridge University Press, 2000.

Struve, Otto, and Velta Zebergs. *Astronomy of the 20th Century.* Macmillan, 1962.

Waller, William H., and Paul W. Hodge. *Galaxies and the Cosmic Frontier.* Harvard University Press, 2003.

ARTICLES AND SCIENCE PAPERS

Abel, Tom, et al. "The Formation of the First Star in the Universe." *Science* 295 (2002): 93–98.

Abraham, Roberto G., et al. "The Gemini Deep Deep Survey. I. Introduction to the Survey, Catalogs, and Composite Spectra." *Astronomical Journal* 127, no. 5 (2004): 2455–83.

Babul, Arif, and Henry C. Ferguson. "Faint Blue Galaxies and the Epoch of Dwarf Galaxy Formation." *Astrophysical Journal* 458 (1996): 100–19.

Bahcall, John N., et al. "What the Longest Exposures from the Hubble Space Telescope Will Reveal." *Science* 248 (1990): 178–83.

Bahcall, N. A., et al. "The Cosmic Triangle: Revealing the State of the Universe." *Science* 284 (1999): 1481–88.

Banks, Thomas. "The Cosmological Constant Problem." *Physics Today* 57 (March 2004): 46–51.

Barkana, Rennan, and Abraham Loeb. "Unusually Large Fluctuations in the Statistics of Galaxy Formation at High Redshift." *Astrophysical Journal* 609, no. 2 (2004): 474–81.

Baron, E., and Simon D. M. White. "The Appearance of Primeval Galaxies." *Astrophysical Journal* 322 (1987): 585–96.

Barton, Elizabeth. "Searching for Star Formation Beyond Reionization." *Astrophysical Journal (Letters)* 604 (2004): L1–4.

Baugh, Carlton, and Carlos Frenk. "How are Galaxies Made?" *Physics Web* (1999), online at http://physicsweb.org/article/world/12/5/91.

Becker, Robert H., et al. "Evidence for Reionization at $z \sim 6$: Detection of a Gunn-Peterson Trough in a $z = 6.28$ Quasar." *Astrophysical Journal* 122 (2001): 2850–57.

Bell, Eric F. "Galaxy Assembly." astro-ph/040823 (2004).

Bell, Eric F., et al. "Nearly 5000 Distant Early-Type Galaxies in COMBO-17: A Red Sequence and Its Evolution Since $z \sim 1$." *Astrophysical Journal* 608, no. 2 (2004): 752–67.

Blitz, Leo. "Global Star Formation From $z = 5 \times 10^{-8}$ to $z = 20$." astro-ph/0501400 (2005).

Bouwens, R. J., et al. "Cloning Hubble Deep Fields. I. A Model-Independent Measurement of Galaxy Evolution." *Astrophysical Journal* 506 (1998): 557–78.

———. "Galaxy Size Evolutions at High Redshift and Surface Brightness Selection Effects Constraints from the Hubble Ultra Deep Field." *Astrophysical Journal (Letters)* 611 (2004): L1–4.

———. "Galaxies at $z \sim 7$–8: z_{850}-Dropouts in the Hubble Ultra Deep Field." *Astrophysical Journal (Letters)* 616, no. 2 (2004): L79–82.

Bromm, Volker, and Richard B. Larson. "The First Stars." *Annual Review of Astronomy and Astrophysics* 42 (2004): 1–38.

Bromm, Volker, and Abraham Loeb. "The Formation of the First Low-Mass Stars from Gas with Low Carbon and Oxygen Abundances." *Nature* 425 (2003): 812–14.

Bromm, Volker, et al. "The First Supernova Explosions in the Universe." *Astrophysical Journal (Letters)* 596 (2003): L135–38.

Bundy, Kevin, et al. "The Mass Assembly Histories of Galaxies of Various Morphologies in the GOODS Fields." astro-ph/0502204 (2005).

Bunker, Andrew, et al. "A Star-Forming Galaxy at $z = 5.78$ in the Chandra Deep Field South." *Monthly Notices of the Royal Astronomical Society* 342 (2003): L47–51.

———. "The Star Formation Rate of the Universe at $z\sim6$ from the Hubble Ultra Deep Field." *Monthly Notices of the Royal Astronomical Society* 355, no. 2 (2004): 374–84.

Caldwell, Robert R. "Dark Energy." *PhysicsWeb* (2004), online at http://physicsweb.org/articles/world/17/5/7.

Carilli, C. L., et al. "Probing the Dark Ages with the Square Kilometer Array." *New Astronomy Reviews* 48, no. 11–12 (2004): 1029–38.

———. "A Search for Dense Molecular Gas in High-Redshift Infrared-Luminous Galaxies." *Astrophysical Journal* 618, no. 2 (2005): 586–91.

Castelvecchi, David. "The Growth of Inflation." *Symmetry* 1 (2005): 13–17.

Cen, Renyue. "The Universe Was Reionized Twice." *Astrophysical Journal* 591 (2003): 12–37.

Chen, Hsiao-Wen. "Discovery of Massive Evolved Galaxies at $z>3$ in the Hubble Ultra Deep Field." *Astrophysical Journal* 615, no. 2 (2004): 603–609.

Cimatti, A., et al. "Old Galaxies in a Young Universe." *Nature* 430 (2004): 184–87.

Conselice, Christopher J. "The Relationship Between Stellar Light Distributions of Galaxies and Their Formation Histories." *Astrophysical Journal* 147 (2003): 1–28.

———. "Unveiling the Formation of Massive Galaxies." *Science* 304 (2004): 399–400.

———. "The Galaxy Structure-Redshift Relationship." astro-ph/0407463 (2004).

Cowen, Ron. "Big Bang Confirmed: Seeing Twists and Turns of Primordial Light." *Science News* 162 (September 28, 2002): 195.

Cuby, Jean-Gabriel. "Discovery of a $z = 6.17$ Galaxy from CFHT and VLT Observations." *Astronomy and Astrophysics* 405 (2003): L19–22.

Davidge, T. J., et al. "Deep ALTAIR+NIRI Imaging of the Disk and Bulge of M31." *Astronomical Journal* 129, no. 1 (2005): 201–19.

Diemand, J., et al. "Earth-mass Dark-matter Haloes as the First Structures in the Early Universe." *Nature* 433 (2005): 389–91.

Djorgovski, S. George. "On the Observability of Primeval Galaxies." *Astronomical Society of the Pacific Conference Series* 24 (1992): 73–91.

——. "The Quest for Protogalaxies." *New Light on Galaxy Evolution*. International Astronomical Union, 2001, 277–86.

——. "Protogalaxies." *Encyclopedia of Astronomy and Astrophysics*. Nature Publishing Group, 2001.

——. "Out of the Dark Ages." *Nature* 427 (2004): 790–91.

Djorgovski, S., et al. "Discovery of a Probable Galaxy with a Redshift of 3.218." *Astrophysical Journal (Letters)* 299 (1985): L1–5.

Egami, E., et al. "Spitzer and Hubble Constraints on the Physical Properties of the $z\sim7$ Galaxy Strongly Lensed by Abell 2218." *Astrophysical Journal (Letters)* 618, no. 1 (2005): L5–8.

Ellis, John, and D. V. Nanopoulos. "Beyond the Standard Model of Cosmology." astro-ph/0411153 (2004).

Ellis, Richard, et al. "A Faint Star-Forming System Viewed Through the Lensing Cluster Abell 2218: First Light at $z \simeq 5.6$?" *Astrophysical Journal (Letters)* 560 (2001): L119–22.

——. "Galaxy Formation and Evolution: Recent Progress." *Galaxies at High Redshift. XI*, Canary Islands Winter School of Astrophysics, Santa Cruz de Tenerife, Spain (2003): 1–28.

Elmegreen, Debra Meloy, et al. "Discovery of Face-On Counterparts of Chain Galaxies in the Tadpole Advanced Camera for Surveys Field." *Astrophysical Journal (Letters)* 604 (2004): L21–23.

Fan, Xiaouhi, et al. "A Survey of $z>5.8$ Quasars in the Sloan Digital Sky Survey. I. Discovery of Three New Quasars and the Spatial Density of Luminous Quasars at $z\sim6$." *Astronomical Journal* 122, no. 6 (2001): 2833–49.

——. "A Survey of $z>5.7$ Quasars in the Sloan Digital Sky Survey II. Discovery of Three Additional Quasars at $z>6$." *Astrophysical Journal* 125 (2003): 1649–59.

Ferguson, Henry C., et al. "The Size Evolution of High-Redshift Galaxies." *Astrophysical Journal (Letters)* 600 (2004): L107–10.

Francis, Paul J., et al. "The Distribution of Lyα-Emitting Galaxies at $z = 2.38$. II. Spectroscopy." *Astrophysical Journal* 614, no. 1 (2004): 75–83.

Freedman, Wendy L., and Michael S. Turner. "Measuring and Understanding the Universe." *Reviews of Modern Physics* 75, no. 4 (2003): 1433–47.

Glazebrook, Karl. "Probing the Redshift Desert Using the Gemini Deep Deep Survey: Observing Galaxy Mass Assembly at $z>1$." *Maps of the Cosmos*, International Astronomical Union, Symposium no. 216 (2003).

Glazebrook, Karl, et al. "A High Abundance of Massive Galaxies 3–6 Billion Years After the Big Bang." *Nature* 430 (2004): 181–83.

Haiman, Zoltán. "Galaxy Formation: Caught in the Act?" *Nature* 430 (2004): 979–80.

Hansen, Brad M. S., et al. "The White Dwarf Cooling Sequence of the Globular Cluster Messier 4." *Astrophysical Journal (Letters)* 574 (2002): L155–58.

Harwit, Martin. "The Growth of Astrophysical Understanding." *Physics Today* (November 2003): 38–43.

Heavens, Alan, et al. "The Star-Formation History of the Universe from the Stellar Populations of Nearby Galaxies." *Nature* 428, no. 6983 (2004): 625–27.

Illingworth, Garth, and Rychard Bouwens. "From $z>6$ to $z\sim2$: Unearthing Galaxies at the Edge of the Dark Ages." astro-ph/0410094 (October 5, 2004).

Irion, Robert. "Surveys Scour the Cosmic Deep." *Science* 303 (2004): 1750–52.

Jimenez, Raul, et al. "Baryonic Conversion Tree: The Global Assembly of Stars and Dark Matter in Galaxies from the SDSS." *Monthly Notices of the Royal Astronomical Society* 356 (2004): 495–501.

Jones, Bernard J. T., Vicent J. Martínez, et al. "Scaling Laws in the Distribution of Galaxies." *Reviews of Modern Physics* 76 (2005): 1211–66.

Kirshner, Robert P. "Throwing Light on Dark Energy." *Science* 300 (2003): 1914–18.

Kneib, Jean-Paul, et al. "A Probable $z\sim7$ Galaxy Strongly Lensed by the Rich Cluster A2218: Exploring the Dark Ages." *Astrophysical Journal* 607 (2004): 697–703.

Krauss, Lawrence M. "Dark Energy and the Hubble Age." *Astrophysical Journal* 604, no. 2 (2004): 481–83.

———. "The Standard Model, Dark Matter, and Dark Energy: From the Sublime to the Ridiculous." astro-ph/0406673 (2004).

Kurk, J. D. "A Lyman α Emitter at $z = 6.5$ Found with Slitless Spectroscopy." *Astronomy and Astrophysics* 422 (2004): L13–17.

Labbé, Ivo. "IRAC Mid-Infrared Imaging of the Hubble Deep Field South: Star Formation Histories and Stellar Masses of Red Galaxies at $z>2$." *Astrophysical Journal* 624 (2005): L81–85.

Lindler, Eric. "On the Trail of Dark Energy." *Cern Courier* 43, no. 7 (September 2003), online at http://www.cerncourier.com/main/article43/7/16.

Loeb, Abraham, and Rennan Barkana. "The Reionization of the Universe by the First Stars and Quasars." *Annual Review of Astronomy and Astrophysics* 39 (2001): 19–66.

Madau, Piero. "The Era of Reionization." *Galaxy Evolution: Theory and Observations.* Revista Mexicana de Astronomía y Astrofísica (Conference Series), 17 (2003): 287–96.

Madau, Piero, and M. Kuhlen. "The Dawn of Galaxies." *Texas in Tuscany. XXI Symposium on Relativistic Astrophysics,* Florence, Italy, December 9–13. World Scientific Publishing, 2003, 31–44.

Madau, Piero, et al. "The Star Formation History of Field Galaxies." *Astrophysical Journal* 498 (1998): 106–16.

Malhotra, S. "An Overdensity of Galaxies at z = 5.9 ± 0.2 in the Ultra Deep Field Confirmed Using the ACS Grism." astro-ph/0501478 (2005).

Martini, Paul. "A Population of Massive Globular Clusters in NGC 5128." *Astrophysical Journal* 610 (2004): 233–46.

McCarthy, Patrick J., et al. "Evolved Galaxies at $z>1.5$ from the Gemini Deep Deep Survey: The Formation Epoch of Massive Stellar Systems." *Astrophysical Journal (Letters)* 614, no. 1 (2004): L9–12.

McIntosh, Daniel H., et al. "The Evolution of Early-Type Red Galaxies with the GEMS Survey: Luminosity-Size and Stellar-Mass Size Since Relations $z\sim1$." astro-ph/0411772 (2004).

Meier, David L. "The Optical Appearance of Model Primeval Galaxies." *Astrophysical Journal* 207 (1976): 343–50.

Miralda-Escudé, Jordi. "The Dark Age of the Universe." *Science* 300 (2003): 1904–1909.

Nomoto, K., et al. "Population III Supernovae and Their Nucleosynthesis." *Memorie della Società Astronomica Italiana* 75 (2004): 312–21.

Ord, Louise M. "From COBE to WMAP: A Decade of Data Under Scrutiny." astro-ph/0412354 (2004).

Ostriker, Jeremiah P., and Paul Steinhardt. "New Light on Dark Matter." *Science* 300 (2003): 1909–1913.

Ostriker, Jeremiah P., and Tarun Souradeep. "The Current Status of Observational Cosmology." *Pramana—Journal of Physics* (2005): 1–13.

Ouchi, Masami. "The Discovery of Primeval Large-scale Structures with Forming Clusters at Redshift 6." *Astrophysical Journal (Letters)* 620, no. 1 (2005): L1–4.

Panagia, Nino. "Detecting Primordial Stars." astro-ph/0410235 (2004).

Partridge, R. B., and P.J.E. Peebles. "Are Young Galaxies Visible?" *Astrophysical Journal* 147 (1967): 868–86.

Peebles, P.J.E. "Probing General Relativity on the Scales of Cosmology." World Scientific Publishing (in press).

Pelló, R., et al. "ISAAC/VLT Observations of a Lensed Galaxy at $z = 10.0$." *Astronomy and Astrophysics* 416 (2004): L35–40.

Perlmutter, Saul. "Supernovae, Dark Energy, and the Accelerating Universe." *Physics Today* (April 2003): 53–60.

Pritchet, C. J. "The Search for Primeval Galaxies." *Astronomical Society of the Pacific* 106 (1994): 1052–67.

Readhead, A.C.S., et al. "Polarization Observations with the Cosmic Microwave Background Imager." *Science* 306 (2004): 836–44.

Rees, Martin J. "Introduction." *Royal Society of London Transactions Series A* 361, no. 1812 (2003): 2427–34.

Riess, Adam G., et al. "Observational Evidence from Supernovae for an Accelerating Universe and a Cosmological Constant." *Astrophysical Journal* 116 (1998): 1009–38.

Schaerer, Daniel. "Towards Observations of the First Stars and Galaxies." *Maps of the Cosmos*, International Astronomical Union, Symposium no. 216 (2003).

———."Expected Properties of Primeval Galaxies and Confrontation with Existing Observations." astro-ph/0309528 (2003).

Schwartz, D. A., and S. N. Virani. "Chandra Measurement of the X-ray Spectrum of a Quasar at $z = 5.99$." *Astrophysical Journal (Letters)* 615 (2004): L21–24.

Schwarzschild, Bertram. "High Redshift Supernovae Reveal an Epoch When Cosmic Expansion Was Slowing Down." *Physics Today* (June 2004): 19–21.

Scott, Douglas. "Cosmic Background Radiation Mini-Review." *Physics Letters* (2004): B. 592.

Shapiro, Paul, et al. "Photoevaporation of Cosmological Minihaloes During Reionization." *Monthly Notices of the Royal Astronomical Society* 348, no. 3 (2004): 753–82.

Shull, J. Michael. "Hot Pursuit of Missing Matter." *Nature* 433 (2005): 465–66.

Somerville, Rachel S., et al. "Cosmic Variance in the Great Observatories Origins Deep Survey." *Astrophysical Journal (Letters)* 600 (2004): L171–74.

Spergel, D. N., et al. "First-Year Wilkinson Microwave Anisotropy Probe (WMAP) Observations: Determination of Cosmological Parameters." *Astrophysical Journal Supplement Series* 148, no. 1 (2003): 175–94.

Spinrad, Hyron. "The Most Distant Galaxies." *Astrophysics Update*. Springer-Praxis Books in Astrophysics and Astronomy, 2003, p. 155.

Springel, Volker, and Lars Hernquist. "The History of Star Formation in a Λ Cold Dark Matter Universe." *Monthly Notices of the Royal Astronomical Society* 339 (2003): 312–34.

Stanway, Elizabeth, et al. "Lyman Break Galaxies and the Star Formation Rate of the Universe at z~6." *Monthly Notice of the Royal Astronomical Society* 342, no. 2 (2003): 439–45.

———. "Hubble Space Telescope Imaging and Keck Spectroscopy of z~6 i-Band Dropout Galaxies in the Advanced Camera for Surveys GOODS Fields." *Astrophysical Journal* 607, no. 2 (2004): 704–20.

Stiavelli, M., et al. "The Possible Detection of Cosmological Reionization Sources." *Astrophysical Journal (Letters)* 610 (2004): L1–4.

———. "Evidence of Primordial Clustering Around the QSO SDSS J1030+0524 at z = 6.28." *Astrophysical Journal (Letters)* 622, no. 1, pt. 2 (2005): L1–4.

Strauss, Michael A. "Reading the Blueprints of Creation." *Scientific American* (February 2004): 54–61.

Taniguchi, Y., et al. "The Subaru Deep-Field Project: Lyman α Emitters at Redshift of 6.6." astro-ph/0407542 (2004).

Tegmark, Max. "Beyond Cosmological Parameters." *American Institute of Physics Conference Proceedings* 666 (2003): 19–32.

Tegmark, Max, et al. "Cosmological Parameters from SDSS and WMAP." *Physical Review D* 69, no. 10 (2004): id. 103501.

Tinsley, B. M. "Accelerating Universe Revisited." *Nature* 273 (1978): 208–11.

———. "Evolutionary Synthesis of the Stellar Population in Elliptical Galaxies. II—Late M Giants and Composition Effects." *Astrophysical Journal* 222 (1978): 14–22.

Treu, Tommaso, et al. "Keck Spectroscopy of Distant GOODS Spheroidal Galaxies Downsizing in a Hierarchical Universe." astro-ph/0502028 (2005).

Tumlinson, Jason, et al. "Nucleosynthesis, Reionization, and the Mass Function of the First Stars." *Astrophysical Journal* 612 (2004): 602–14.

Tyson, J. A. "Deep CCD Survey: Galaxy Luminosity and Color Evolution." *Astronomical Journal* 96 (1988): 1–23.

Venkatesan, Aparna, et al. "Evolving Spectra of Population III Stars: Consequences for Cosmological Reionization." *Astrophysical Journal* 584 (2003): 621–32.

Vestergaard, M. "Early Growth and Efficient Accretion of Massive Black Holes at High Redshift." *Astrophysical Journal* 601, no. 2 (2004): 676–91.

Walter, Fabian. "Resolved Molecular Gas in a Quasar Host Galaxy at Redshift $z = 6.42$." *Astrophysical Journal (Letters)* 615, no. 1 (2004): L17–20.

Wang, J. X., et al. "An Overdensity of Lyman-α Emitters at Redshift $z{\sim}5.7$ Near the Hubble Ultra Deep Field." astro-ph/0501479 (2005).

Wilkinson, P. N., et al. "The Exploration of the Unknown." *New Astronomy Reviews* 48, no. 11–12 (2004): 1551–63.

Wirth, Gregory D. "Old Before Their Time." *Nature* (2004): 149–50.

Wyithe, J. Stuart B., and Abraham Loeb. "Was the Universe Reionized by Massive Metal-free Stars?" *Astrophysical Journal (Letters)* 588, no. 2 (2003): L69–72.

———. "A Characteristic Size of ~10 Mpc for the Ionized Bubbles at the End of Cosmic Reionization." *Nature* 432 (2004): 194–96.

Yan, Haojing, and Rogier A. Windhorst. "Candidates of $z \simeq 5.5$–7 Galaxies in the HST Ultra Deep Field." *Astrophysical Journal (Letters)* 612, no. 2 (2004): L93–96.

Yoshida, Naoki, et al. "The Era of Massive Population III Stars: Cosmological Implications and Self-Termination." *Astrophysical Journal* 605 (2004): 579–90.

Zwicky, Fritz. "Nebulae as Gravitational Lenses." *Physical Review* 51 (1937): 290.

ACKNOWLEDGMENTS

If I occasionally sound like I know what I'm talking about in this book, it's because I have had access to good—no, *great*—sources of information. In this business you can't be any better than your sources, at least the ones who deign to speak with you at length, even when you're not a card-carrying astrophysicist.

So first and foremost, I want to thank the many scientists who returned my phone calls and emails, or who were generous enough to grant me an interview or just chat. Most notably these include Roberto Abraham, John Bahcall, Neta Bahcall, Elizabeth Barton, Steven Beckwith, Eric Bell, Rychard Bouwens, Andy Bunker, Chris Carilli, Christopher Conselice, Jean-Gabriel Cuby, George Djorgovski, Simon Driver, Richard Ellis, Owen Gingerich, Bradley Hansen, Gary Hinshaw, Raul Jimenez, Shardha Jogee, Bill Keel, Ken Kellermann, Edward Kolb, Mario Livio, Jim Lyke, Barry Madore, Paul Martini, Ann Nelson, Ken'ichi Nomoto, P.J.E. Peebles, James Rhodes, Adam Riess, Michael Santos, Paul Shapiro, Michael Shull, J. D. Smith, Rachel Somerville, Daniel Stark, Max Tegmark, David Thompson, Rodger Thompson, Jason Tumlinson, Liese van Zee, Aparna Venkatesan, J. Craig Wheeler,

Robert Williams, and Rogier Windhorst. My apologies to anyone I may have inadvertently left out.

I wish to offer my sincere appreciation to Stephen P. Maran, chief press officer of the American Astronomical Society, who arranged for interview space at the AAS meetings, helped line up interviewees, and has always been at the ready with sage advice and support. I'm beholden to the staffs at the Gemini and Keck observatories for allowing us very generous access to their facilities on Mauna Kea. A special thanks to Peter Michaud for being our guide and resident expert at Gemini, and to Rich Matsuda for spontaneously volunteering his services as guide at Keck.

I shudder to think where any science writer would be without the NASA/IPAC Extragalactic Database (http://nedwww.ipac.caltech.edu/), the astro-ph site maintained by Cornell University (http://xxx.lanl.gov/list/astro-ph/new), and the NASA Astrophysics Data System (http://adswww.harvard.edu/). A special thanks to Dr. Edward Wright's JavaScript Cosmology Calculator (http://www.astro.ucla.edu/~wright/CosmoCalc.html), where you, too, can be a cosmologist.

In writing about something as recondite as cosmology, it's invaluable to have a sympathetic inside expert, someone you can seek out when you have the inevitable moronic questions that you're too embarrassed to ask anyone else. For me, that person was George Djorgovski. Many of his insightful comments found their way into these pages. I don't think he will mind my referring to him as one of the true natural philosophers working in astronomy today.

My heartfelt thanks to Regula Noetzli, who championed my book proposal above and beyond the calling of any literary agent, and to my editor, Joseph Wisnovsky, whose name opened many doors and whose fine-point editing kept me on the beam.

ACKNOWLEDGMENTS

Finally, I'd like to thank Alexandra Witze, my award-winning science-writing wife, for reading each chapter (despite her penchant for planetary geology), making her usual astute editorial suggestions, being a patient sounding board, and most of all encouraging me at a time when I needed it most.

INDEX

159, 172, 173; abnormal galaxies in, 134; cosmological principle and, 63–64; "fossil light" in, 55; mergers of galaxies in, 136; primordial objects in, 182
Hubble Deep Field South, 6, 64, 182
Hubble expansion, 83
Hubble Higher-z Supernova Research Team, 95
Hubble Space Telescope, 5–6, 38, 51, 56, 98, 117, 148–50, 166–67; acuity of, 19; Advanced Camera for Surveys (ACS), 42, 45, 176; deep field exposures of, 17, 64, 138 (see also Hubble Deep Field; Hubble Ultra Deep Field); in Great Observatories Origins Deep Survey, 34; Near Infrared Camera and Multi-Object Spectrometer (NICMOS), 42–46, 49, 50, 53, 57, 77; operating costs of, 109; successors to, 171–74
Hubble Treasury Program, 45
Hubble Ultra Deep Field (HUDF), 4–6, 22–23, 34, 36, 46n, 49, 53, 79, 158, 159, 172, 181; abnormal galaxies in, 134; dwarf galaxies in, 140; "fossil light" in, 55; location of image frames of, 37; narrow width of, 56–57; primordial objects in, 182; public release of, 4, 21–22, 41–43, 45; unbiased spectroscopy of, 33
hydrogen, 9, 74, 75, 80, 85, 93; reionization and, 34–37, 47–49, 125
hypernovae, 74, 175, 181–82

Indiana University, 85
inflation, 68–71, 90–91, 103, 169–70
infrared telescopes, 42–43, 66
Institute for Advanced Study (Princeton), 49, 67
International SKA Steering Committee, 179
ionization, 47, 48, 75, 81; see also reionization
Italian National Institute of Nuclear Physics, 103

James Webb Space Telescope (JWST), 74, 86, 138, 172, 174–76, 180
Jansky, Karl, 120, 178
Jimenez, Raul, 155–57
Jodrell Bank Observatory, 178
Jogee, Shardha, 57
Joint Dark Energy Mission (JDEM), 167

Keck Observatory, 20, 109–27
Kellermann, Ken, 179
Kneib, Jean-Paul, 37
Kolb, Edward W., 103

Laboratory for Underground Nuclear Physics, 87
lambda force, 58
Large Hadron Collider, 170
Large Magellanic Cloud, 129, 151
Larkin, James, 154
last scattering surface, 73–74
Lemaître, Georges, 60–61, 64